2005

THE STATE OF FOOD AND AGRICULTURE

FAO Agriculture Series No. 36

ISSN 0081-4539

FOOD AND AGRICULTURE ORGANIZATION OF THE UNITED NATIONS
Rome, 2005

Produced by the
Editorial Production and Design Group
Publishing Management Service
FAO

The designations employed and the presentation of material in this information product do not imply the expression of any opinion whatsoever on the part of the Food and Agriculture Organization of the United Nations concerning the legal or development status of any country, territory, city or area or of its authorities, or concerning the delimitation of its frontiers or boundaries. The mention or omission of specific companies, their products or brand names does not imply any endorsement or judgement by the Food and Agriculture Organization of the United Nations.

ISBN 92-5-105349-9

All rights reserved. Reproduction and dissemination of material in this information product for educational or other non-commercial purposes are authorized without any prior written permission from the copyright holders provided the source is fully acknowledged. Reproduction of material in this information product for resale or other commercial purposes is prohibited without written permission of the copyright holders. Applications for such permission should be addressed to:

Chief
Publishing Management Service
Information Division
FAO
Viale delle Terme di Caracalla, 00100 Rome, Italy
or by e-mail to:
copyright@fao.org

© FAO 2005

Contents

Foreword	vii
Acknowledgements	x
Glossary	xi
Explanatory note	xiii

PART I
Agricultural trade and poverty: can trade work for the poor?

1.	**Introduction and overview**	**3**
	Trade, poverty and food security: what are the linkages?	6
	Overview of the report	7
2.	**Trends and patterns in international agricultural trade**	**12**
	Agricultural trade and the world economy	12
	The shifting geography of agricultural trade	14
	Agricultural trade in the least developed countries	16
	Agricultural trade within regions	17
	Processed products and the role of supermarkets	20
	Key findings	25
3.	**The agricultural trade policy landscape**	**26**
	Evolution of agricultural trade policy	26
	Domestic support	30
	Export competition	33
	Market access	38
	Key findings	47
4.	**Macroeconomic impacts of agricultural trade reform**	**48**
	Modelling trade policy reform	48
	Computable general equilibrium model results	50
	Agriculture sector model results	57
	Key findings	58
5.	**Poverty impacts of agricultural trade reforms**	**60**
	Agriculture's role in poverty reduction	60
	Trade's role in poverty reduction	63
	Agricultural trade reform and poverty	66
	Impact of trade reforms on factor markets	72
	Trade reforms, productivity and economic growth	74
	Implications for policy research	78
	Key findings	79
6.	**Trade and food security**	**80**
	What is food security?	80
	Correlations between trade and hunger	82
	Trade liberalization and food security	84
	Case studies in macroeconomic and trade reforms	86
	Key findings	96
7.	**Making trade work for the poor: the twin track approach to hunger and poverty reduction**	**98**
	Track one: creating opportunities	99
	Track two: ensuring access	102
	Are we on the right path?	103
	Summary	105

Part II
World and regional review: facts and figures

1.	Trends in undernourishment	117
2.	Food emergencies and food aid	119
3.	External assistance to agriculture	121
4.	Crop and livestock production	123
5.	World cereal supply situation	126
6.	International commodity price trends	127
7.	Agricultural trade	130
8.	Fisheries: production, utilization and trade	134
9.	Forestry	137

Part III
Statistical annex

Notes on the annex tables		143
Table A1	Countries and territories used for statistical purposes in this publication	148
Table A2	Food security and nutrition	150
Table A3	Agricultural production and productivity	154
Table A4	Population and labour force indicators	159
Table A5	Land use	165
Table A6	Trade indicators	172
Table A7	Economic indicators	177
Table A8	Total factor productivity	182

References	187
Special chapters of *The State of Food and Agriculture*	193
Selected publications	195

SPECIAL CONTRIBUTION

Can trade work for the poor? A view from civil society   108

TABLES

1. Destination of agricultural exports by region   22
2. Origin of agricultural imports by region   23
3. OECD producer support estimate   30
4. Measures of domestic support   31
5. Country-level agricultural tariff data, 2000–02   43
6. Welfare gains from CGE studies of trade liberalization   52
7. Bilateral trade: percentage change in value of bilateral import volumes   55
8. Effects of trade liberalization on unskilled wages by sector and scenario   56
9. Impacts of policy reform on world commodity prices   57
10. Food and hunger indicators by region   80
11. Average applied and bound MFN tariffs   88
12. Ratio of total value of food imports to total value of agricultural exports   92
13. Changes in the proportion of the population undernourished, food production, rural poverty and economic growth   93
14. Per capita availability of calories and protein, 1980/82–1999/2001   95
15. Per capita shipments of food aid in cereals   120

BOXES

1. What other multilateral agencies conclude about trade and development   4
2. Main provisions of the Uruguay Round Agreement on Agriculture   28
3. The European Union's tariff quota regime for dairy products   38
4. Tariffs as tax revenue   41
5. Key features of computable general equilibrium models   50
6. What do we know about poverty reduction?   61
7. Agricultural households   68
8. Impact of agricultural liberalization on poverty in Brazil   70
9. Why trade matters for reducing poverty and improving food security   76
10. Cashew market liberalization in Mozambique   85
11. Breaking the cycle of hunger and poverty: a twin track strategy to reduce hunger and poverty   99

FIGURES

1. Growth in global GDP and global trade in goods and services   13
2. Growth in global agricultural GDP and global trade in agricultural goods   13
3. Ratio of trade to GDP in the global economy   13
4. World agricultural exports: total and as share of merchandise exports   14
5. Share of developing countries in agricultural and total merchandise trade   15
6. Agricultural trade in developed and developing countries   16
7. Ratio of trade to GDP in developed and developing countries   17
8. Agricultural trade in the developing country regions   18
9. Regional shares in world agricultural trade   19
10. Agricultural trade in the least developed countries   20
11. Ratio of trade to GDP in the least developed countries   20
12. Share of processed products in agricultural exports   21
13. Subsidized exports as share of total EU exports of selected products, 1995–2001   34
14. Tariff escalation for fibres, textiles and clothing   42
15. Agricultural GDP and undernourishment, 1998–2002   62
16. Agricultural employment and undernourishment, 1998–2002   62

17.	Agricultural trade and undernourishment, 1998–2002	64
18.	Agricultural exports and undernourishment, 1998–2002	64
19.	Agricultural imports and undernourishment, 1998–2002	65
20.	Food imports and undernourishment, 1998–2002	65
21.	Integration of agriculture into world markets and undernourishment, 1998–2002	65
22.	Regional impact of trade liberalization in Mexico	67
23.	Initial impact of WTO accession on rural and urban household real income in China	69
24.	Impact of MERCOSUR on household real income in Argentina	73
25.	Impact of trade liberalization on household real income in Mexico	73
26.	Conceptual framework for food insecurity	81
27.	Percentage undernourished plotted against ratio of agricultural trade to agricultural GDP	82
28.	Percentage underweight plotted against ratio of agricultural trade to agricultural GDP	83
29.	Reform–response–result framework	84
30.	Evolution of real domestic prices and the real effective exchange rate in Chile	90
31.	Evolution of real domestic prices and the real effective exchange rate in Ghana	91
32.	Change in average food availability vs change in undernutrition prevalence during the 1990s	94
33.	Agricultural capital stock per agricultural worker in developing countries by prevalence of undernourishment in 2000–2002	104
34.	Agricultural orientation of public investment	105
35.	Long-term trend in external assistance to agriculture, 1974–2002	105
36.	External assistance to agriculture per agricultural worker by prevalence of undernourishment, 1998–2000	106
37.	Undernourished population by region, 2000–02	117
38.	Trend in number of undernourished in developing countries, by region	118
39.	Trend in percentage of undernourished in developing countries, by region	118
40.	Recipients of food aid in cereals	119
41.	Recipients of food aid in non-cereals	121
42.	Commitments of external assistance to agriculture, by main recipient regions	122
43.	External assistance to agriculture per agricultural worker	122
44.	Changes in crop and livestock production	124
45.	Long-term trend in per capita food production by region and country group	125
46.	World cereal production and utilization	126
47.	World cereal stocks and stocks-to-utilization ratio	127
48.	Commodity price trends	128
49.	Annual change in value of global agricultural exports	130
50.	Global agricultural exports	131
51.	Agricultural imports and exports, by region	131
52.	World fish production, China and rest of the world	135
53.	Trade in fish and fish products, developed and developing countries	136
54.	Net exports of fish and fish products and selected agricultural commodities in developing countries	136
55.	World roundwood production	137
56.	Production, consumption, imports and exports of industrial roundwood in 2002	138
57.	Roundwood production, developed and developing countries	138
58.	Value of trade in forest products	139
59.	Industrial roundwood production by region, 2002	140

Foreword

The State of Food and Agriculture 2005 examines the linkages among agriculture, trade and poverty and asks whether international agricultural trade, and its further reform, can help overcome extreme poverty and hunger.

The global statistics on poverty and hunger are all too familiar. An estimated 1.2 billion people live on less than one dollar a day and FAO's most recent estimates indicate that 852 million people lack sufficient food for an active and healthy life. There is now also an increased awareness that extreme poverty and hunger are largely rural phenomena. Most of the world's poor and hungry people live in rural areas and depend on agriculture for their livelihoods. To the extent that agriculture is affected by trade, trade will necessarily affect the livelihoods and food security of the world's most vulnerable people.

The global economy is becoming increasingly integrated through trade, and agriculture is part of this larger trend. For some countries, agricultural trade expansion – sparked by agricultural and trade policy reforms – has contributed to a period of rapid pro-poor economic growth. Indeed, some of the countries that have been most successful in reducing hunger and extreme poverty have relied on trade in agricultural products, either exports or imports or both, as an essential element of their development strategy.

Many of the poorest countries however, have not had the same positive experience. Rather, they are becoming more marginalized and vulnerable, depending on imports for a rising share of their food needs without being able to expand and diversify their agricultural or non-agricultural exports. FAO believes that the reform process under way must consider the specific circumstances of these countries, particularly their stage of agricultural development and the complementary policies needed to ensure their successful integration into global agricultural markets.

FAO has long recognized that agricultural trade is vital for food security, poverty alleviation and economic growth. Food imports are a fundamental means of supplementing local production in ensuring the provision of minimum supplies of basic foodstuffs in many countries. Agricultural exports are an important source of foreign exchange earnings and rural income in many developing countries. Reducing trade-distorting agricultural subsidies and barriers to agricultural trade can serve as a catalyst for growth as producers worldwide could then compete on the basis of their comparative advantage.

However, international trade in agricultural products is characterized by a number of problems that do not allow competition on the basis of comparative advantage. The markets for many temperate-zone products and basic food commodities are substantially distorted by government subsidies and protection, particularly in Organisation for Economic Co-operation and Development (OECD) countries. Some developed countries continue to subsidize their farmers and, where this leads to market surpluses, even their agricultural exports. For other agricultural products, particularly tropical ones such as coffee, tea, natural fibres, tropical fruits and vegetables, the problems include high as well as complex and seasonal tariffs and significant tariff escalation.

These market distortions tend to lower world market prices for basic foodstuffs and limit market access. This has helped net food-importing low-income countries to keep their food import bills low, but has also sent wrong signals to the governments of developing countries that have sometimes misled them to neglect their own agriculture. Low prices and lack of investment have hindered agricultural and rural development in poor countries. In this context it must be emphasized that it is in the developed countries' interests that developing countries grow faster, not least

because such growth would increase the size of markets for developed country non-agricultural exports.

The developing countries too have important decisions to make. Some developing country exporters would benefit from the liberalization of OECD agricultural policies, but benefits for developing countries are also expected to result from liberalization of trade among them. Indeed, many benefits from global agricultural trade liberalization for developing countries would be the result of their own policy reforms. South–South agricultural trade is expanding rapidly as incomes rise, cities grow and lifestyles shift towards more diverse diets. These are the growth markets of the future.

It should be noted, however, that some developing countries may not gain from further agricultural trade liberalization. Some countries that depend on preferential access to protected OECD markets for their agricultural exports would lose if those preferences were eroded. Net food importing countries would also be harmed, especially in the short run, in so far as the removal of OECD subsidies would lead to higher prices of basic foodstuffs on world markets.

Although there seems to be broad consensus that trade liberalization fosters efficiency and economic growth, the immediate results for the poor and food-insecure seem to be mixed in the present situation of distorted agricultural commodity markets. Experience shows that gains and losses and the distribution of winners and losers among individuals and countries are determined by context. In practice, a great deal seems to depend on the existence of complementary factors. International trade and trade liberalization can best promote sustainable reductions in hunger and poverty if appropriate complementary measures are put in place.

These measures include, on the one hand, investments that would enable people to take advantage of the opportunities presented by trade and, on the other hand, social safety nets to ensure that the weakest and most vulnerable members of society are protected from the potential disruptions that arise from trade reform. We must always pay particular attention to the specific difficulties faced by the least-developed countries, the low-income food-deficit countries and other vulnerable groups.

Among the most important of these investments are measures such as reducing the large variations in agricultural production in rain-dependent areas through small-scale water projects implemented at the village and community levels; improving rural roads so that inputs can more easily reach the producers and production the markets; and improving all components of the marketing chain. Especially needed are better storage and packaging facilities at the farm level and throughout the marketing process, as well as market facilities, slaughterhouses and ports. Equally important is investment in capacity building to enable countries to comply with quality and safety standards and with the World Trade Organization Agreements on Sanitary and Phytosanitary Measures and Technical Barriers to Trade; this includes the provision of skills training, equipment and resources, and strengthening of institutions to facilitate countries' active participation in standard-setting bodies.

Such investment in agriculture and rural areas has multiple payoffs, not the least of which is the increased capacity of developing countries to become more effective participants in the international economy. With proper assistance from wealthier countries, trade standards can be transformed from a threat to an opportunity.

FAO's ongoing studies and analyses do provide encouraging lessons and overall policy guidance. Among these many important lessons is the need for policy-makers to consider more carefully than they have in the past how trade policies can be used positively to promote pro-poor growth. This involves actively implementing policies and making investments that complement trade reforms to enable the poor to take advantage of trade-related opportunities, while establishing safety nets to protect vulnerable members of society.

The Millennium Declaration underscores the importance of international trade in the context of development and the elimination of poverty. In the Millennium Declaration, governments committed themselves, *inter alia*, to the creation of an open, equitable, rule-based and non-discriminatory multilateral trading system. Such a system is essential if international agricultural trade is to promote more equitable economic growth and contribute to the goals of poverty alleviation and food security.

Jacques Diouf
FAO DIRECTOR-GENERAL

Acknowledgements

The State of Food and Agriculture is prepared by FAO's Agricultural and Development Economics Division. The team is led by Terri Raney, Senior Economist and Editor, and includes André Croppenstedt, Annelies Deuss, Jakob Skoet and Slobodanka Teodosijevic. Stella di Lorenzo and Paola di Santo provide secretarial and administrative support. Randy Stringer, Chief, Comparative Agricultural Development Service, and Prabhu Pingali, Director, Agricultural and Development Economics Division, provide overall supervision and guidance.

Part I, "Agricultural trade and poverty, can trade work for the poor?" was prepared under the direction of Joe Francois, Randy Stringer and Alexander Sarris. Numerous people from several FAO technical units, in particular the Commodities and Trade Division, as well as other international organizations and independent experts provided helpful advice, assistance and guidance. The key background research for Part I is based on work by Joe Francois, Tom Hertel, Phillip Killicoat, Maros Ivanic, Will Martin, Randy Stringer, Jacob Skoet, Frank van Tongeren and Wang Zhi. The report benefited greatly from critical comments, analytical reviews and substantive editing by Kym Anderson, Jelle Bruinsma, Walter Falcon, Hartwig de Haen, Tim Josling, Jamie Morrison, Prabhu Pingali, Ramesh Sharma and Alberto Valdés. Colleagues in FAO's Statistics Division provided data and related statistical inputs.

Chapter 1 (Introduction and overview). Draft text was provided by Joe Francois, Tim Josling, Terri Raney and Randy Stringer.

Chapter 2 (Trends and patterns in international agricultural trade). Text was prepared by Annelies Deuss, Jacob Skoet and Randy Stringer.

Chapter 3 (The agricultural trade policy landscape). Contributors include Joe Francois, Tim Josling, Will Martin, Jakob Skoet, Frank van Tongeren and Wang Zhi.

Chapter 4 (Macroeconomic impacts of agricultural trade reform). This chapter is based on a background paper prepared by Joe Francois and Frank van Tongeren. Tom Hertel, Terri Raney and Jakob Skoet provided additional contributions.

Chapter 5 (Poverty impacts of agricultural trade reforms). This chapter is based on a background paper prepared by Tom Hertel and Maros Ivanic. Additional inputs were provided by Randy Stringer and Alberto Valdés.

Chapter 6 (Trade and food security). Phillip Killicoat, Annelies Deuss, Terri Raney and Jakob Skoet prepared this chapter, which is based in large part on research by the FAO Commodities and Trade Division. This research was presented to the FAO Committee on Commodity Problems in May 2005 in the document *Food security in the context of the context of economic and trade policy reforms: Insights from country experiences* (CCP 05/11).

Chapter 7 (Making trade work for the poor: the twin track approach to hunger and poverty reduction). This concluding chapter was prepared by Annelies Deuss, Philip Killacoat, Prabhu Pingali, Terri Raney, Randy Stringer and Jakob Skoet.

Part II, "World and regional review: facts and figures", was prepared by André Croppenstedt, Annelies Deuss and Jakob Skoet.

Part III, "Statistical annex", was prepared by André Croppenstedt, Annelies Deuss and Terri Raney.

The team is particularly grateful to the *State of Food and Agriculture* External Advisory Board, comprising Walter Falcon (Chair), Bina Agarwal, Kym Anderson, Simeon Ehui, Franz Heidhues and Eugenia Muchnik, who provided valuable guidance on the scope and focus of the report.

The report benefited from the work of the editors, designers and layout artists of the FAO Publishing Management Service.

Glossary

AMS	aggregate measurement of support
AoA	(Uruguay Round) Agreement on Agriculture
CAP	Common Agricultural Policy
CGE	computable general equilibrium
CSO	civil society organization
CV	coefficient of variation
EAA	external assistance to agriculture
EPA	United States Environmental Protection Agency
EU	European Union
EV	equivalent variation
FDI	foreign direct investment
GATT	General Agreement on Tariffs and Trade
GDP	gross domestic product
GTAP	Global Trade Analysis Project
IMF	International Monetary Fund
IPC	International NGO/CSO Planning Committee for Food Sovereignty
LDC	least developed country
MDG	Millennium Development Goal
MFN	most-favoured nation
NAFTA	North American Free Trade Agreement
NFIDC	net food-importing developing country
NGO	non-governmental organization
NTB	non-tariff barrier
OECD	Organisation for Economic Co-operation and Development
PPP	purchasing power parity

PRSP	Poverty Reduction Strategy Paper
PSE	producer support estimate
ROA	Roles of Agriculture Research Project (FAO)
SSG	special safeguard (mechanism)
STE	state trading enterprise
TRQ	tariff rate quota
UNCLOS	United Nations Convention on the Law of the Sea
UNCTAD	United Nations Conference on Trade and Development
UNDP	United Nations Development Programme
USDA	United States Department of Agriculture
WTO	World Trade Organization

Explanatory note

The statistical information in this issue of *The State of Food and Agriculture* has been prepared from information available to FAO up to November 2005.

Symbols
The following symbols are used:
- – = none or negligible (in tables)
- ... = not available (in tables)
- $ = US dollars

Dates and units
The following forms are used to denote years or groups of years:
2003/04 = a crop, marketing or fiscal year running from one calendar year to the next
2003–04 = the average for the two calendar years
Unless otherwise indicated, the metric system is used in this publication.
"Billion" = 1 000 million.

Statistics
Figures in statistical tables may not add up because of rounding. Annual changes and rates of change have been calculated from unrounded figures.

Production indices
The FAO indices of agricultural production show the relative level of the aggregate volume of agricultural production for each year in comparison with the base period 1989–91. They are based on the sum of price-weighted quantities of different agricultural commodities after the quantities used as seed and feed (similarly weighted) have been deducted. The resulting aggregate therefore represents disposable production for any use except seed and feed.

All the indices, whether at the country, regional or world level, are calculated by the Laspeyres formula. Production quantities of each commodity are weighted by 1989–91 average international commodity prices and summed for each year. To obtain the index, the aggregate for a given year is divided by the average aggregate for the base period 1989–91.

Trade indices
The indices of trade in agricultural products are also based on the base period 1989–91. They include all the commodities and countries shown in the *FAO Trade Yearbook*. Indices of total food products include those edible products generally classified as "food".

All indices represent changes in current values of exports (free on board [f.o.b.]), and imports (cost, insurance, freight [c.i.f.]), expressed in US dollars. When countries report imports valued at f.o.b., these are adjusted to approximate c.i.f. values.

Volumes and unit value indices represent the changes in the price-weighted sum of quantities and of the quantity-weighted unit values of products traded between countries. The weights are, respectively, the price and quantity averages of 1989-91 which is the base reference period used for all the index number series currently computed by FAO. The Laspeyres formula is used to construct the index numbers.

Explanatory note

Part I

AGRICULTURAL TRADE AND POVERTY
Can trade work for the poor?

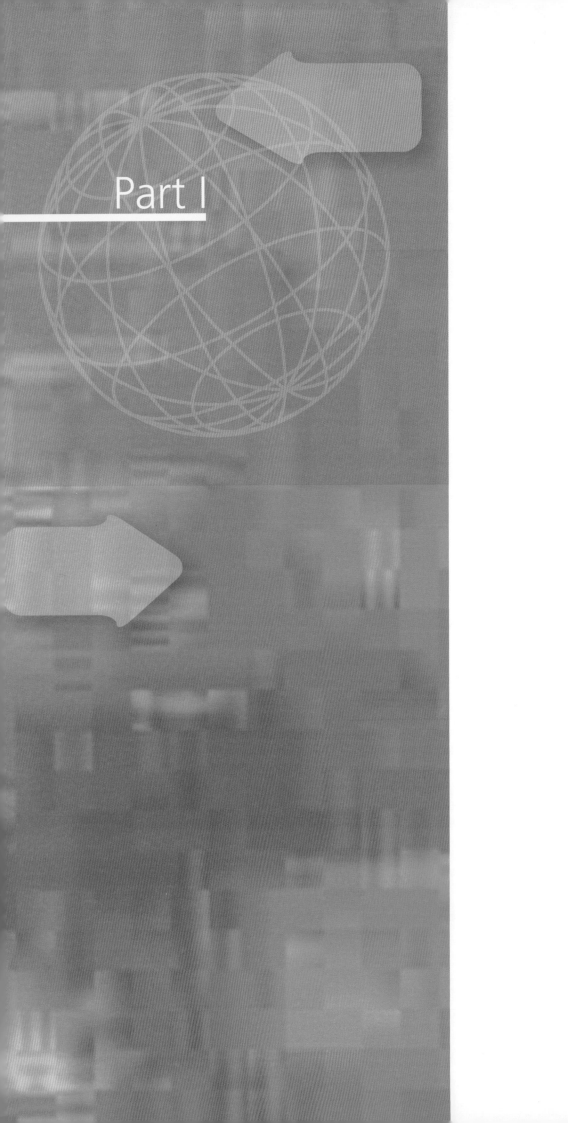

Part I

1. Introduction and overview

The debate over the role of trade in economic growth and poverty reduction has a long history. This often contentious debate dates back more than 50 years at FAO and lies at the very roots of economics.

Advocates of freer trade argue that trade promotes growth and that growth reduces poverty. This view maintains that trade barriers such as import tariffs and subsidies generally benefit a powerful, protected few at the expense of the many. Reducing trade barriers promotes more efficient resource use.

Greater efficiency means that societies can produce more of the things people want, within their limited resources, raising overall social welfare. The poor are able to improve their levels of nutrition, health and education, creating a virtuous circle of rising productivity and poverty reduction.

Critics of freer trade argue that this "neoclassical" model is flawed and that it fails to account adequately for market imperfections and for inequitable power relations that govern the multilateral trade negotiation process. Trade liberalization damages food security, they argue, because liberalization benefits only the larger and more export-oriented farmers, leads to scale incentives and size concentration, marginalizes small farmers and creates unemployment and poverty.

Critics also maintain that trade liberalization holds no guarantee that everyone will benefit, even in the long run, arguing that in reality it is the poorest and vulnerable members of society who suffer most from the market disruptions arising from the reform process.

They claim, moreover, that agricultural imports from developed countries undermine the economic and social fabric of poor rural areas, stalling the traditional engine of growth in agrarian societies. Their fear is that the more the developing countries open their borders, the more they expose poor food consumers to price shocks and small food producers to risks and disincentives.

Pointing to the existing international trading system for agriculture, many criticize the import barriers, export subsidies and domestic support retained by some industrial countries in spite of recent progress under the World Trade Organization (WTO) Agreement on Agriculture. They question how farmers in developing countries can compete when their governments had already agreed to trade and agricultural policy changes promoted by the World Bank and the International Monetary Fund (IMF) under structural adjustment programmes.

Advocates of agricultural trade liberalization argue, on the other hand, that this view is too pessimistic and one-sided, and that the adjustments associated with policy reform are temporary and the efficiency gains from trade outweigh these transitory costs. They claim that trade barriers are a costly and ineffective way of supporting food security and agricultural development in poor countries. Rather, productivity-enhancing investments in market institutions, infrastructure, technology and human capital represent a better strategy for pro-poor growth.

While recognizing the imperfect nature of the WTO trade reform process, supporters

> **BOX 1**
> **What other multilateral agencies conclude about trade and development**
>
> Four recent reports on trade and development highlight the importance the international community places on the promise of trade. In common with *The State of Food and Agriculture 2005*, these other UN agencies all call for: (i) an end to OECD countries supporting their agriculture sectors in ways that harm developing countries; (ii) more effective approaches to the risks caused by negative commodity price shocks; (iii) more effective market access for developing countries; and (iv) enhanced South–South cooperation in the field of trade and investment.
>
> Each agency arrived at the common conclusions presented above even though they focused on different trade- and development-related themes. For example, the WTO's *World Trade Report 2004* examines the impact of domestic policies on trade, arguing that the benefits from good trade policy may be undermined if governments do not also pursue appropriate domestic sector-specific policies. While trade policy can have a positive impact on a country's growth and development prospects, the WTO report stresses the importance of ensuring coherence in policy formulation, pointing out that inconsistencies in policy stances or neglect in particular areas can diminish valuable trading opportunities.
>
> UNCTAD's *Least Developed Countries Report* for 2004 assesses the relationship between international trade and poverty within the least developed countries (LDCs), concluding that international trade has not fulfilled its major potential role in poverty reduction in LDCs. Reasons include weak trade performance, weaker linkages between trade and economic growth than in the more advanced developing countries and a tendency for export expansion in very poor economies to be associated with an exclusionary rather than inclusive form of economic growth.
>
> The World Bank's *Global Economic Prospects* for 2004 concentrates heavily on the international trading regime and its implications for developing countries. The report argues the case for trade liberalization and the positive impact on developing countries, illustrating a

argue that the situation for developing countries could have been much worse without the disciplines of the Agreement on Agriculture. They point to the "subsidy wars" of the mid-1980s that generated huge surplus stocks in Europe and North America, severely depressing and destabilizing global commodity prices. WTO disciplines helped reduce these excesses and may have prevented far worse.

They maintain that the structural adjustment programmes implemented by many developing countries in the 1980s and 1990s were essential in order to correct unsustainable budget deficits and overvalued currencies. To the extent that structural adjustment reforms have actually been implemented – and the experts are divided sharply on this question – the severe "urban bias" that penalized agriculture in many developing countries has been reduced.

It took 50 years of successive multilateral trade negotiations to bring down tariffs on industrial goods. Supporters say that the process has just begun for agriculture and further real reform is needed, but if governments manage the adjustment properly within the broad policy latitude they retain under the WTO, opportunities will open up for those displaced by competition from imports.

So, which story best fits reality? Does agricultural trade liberalization condemn agrarian societies to remain in poverty? Is the improved economic efficiency that comes from trade liberalization enough to offset job and income losses experienced by vulnerable groups and individuals? How are poverty and food security affected as borders open up? Are development policies evolving in ways that take best advantage of emerging trade opportunities?

Are the institutions, infrastructure and safety net programmes available in developing countries sufficient to cope pro-poor programme of trade liberalization in all sectors, which, if implemented over five years to 2010 could produce gains for developing countries of nearly $350 billion by 2015 and reduce poverty by 8 percent. This report contains a chapter devoted specifically to agricultural trade, providing a detailed analysis of trends and patterns in world agricultural trade and of patterns of agricultural protection, and a review of reform proposals in the Doha Round. In its summary, the chapter lays out the main components of a pro-poor agenda for policy change in agricultural trade.

UNDP's *Making global trade work for people* (2003) concludes that trade should be seen as a means to development rather than an end. Trade has enormous potential to contribute to human development, yet the current system has fallen far short of expectations and its many inequities are at the core of continuing controversies surrounding economic globalization. Among the key lessons, the report highlights the experiences of industrial countries and successful developing countries. First, economic integration with the world economy is an outcome of growth and development, not a prerequisite. Second, institutional innovations – many of them unorthodox and requiring considerable domestic policy space and flexibility – have been crucial for successful development strategies and outcomes.

Finally, the UNDP report argues that the design of the multilateral trade regime needs to shift from one based on a market access perspective to one based on a human development perspective. It should also be evaluated not on the basis of whether it maximizes the flow of goods and services but on whether trade arrangements – current and proposed – maximize possibilities for human development, especially in developing countries.

Source: WTO, 2004a; UNCTAD, 2004; World Bank, 2003 and UNDP, 2003.

with the risks to vulnerable groups? How can developing countries compete with the economic and political clout of the much larger and much richer industrial countries? Can freer trade help overcome the mismatch between abundant global food supplies and starving families?

Can trade work for the poor? This is the key question that this year's *State of Food and Agriculture* addresses. It is also a key question for the international development community. The United Nations Millennium Declaration underscores the importance of international trade in the context of development and the elimination of poverty.[1] In the Millennium Declaration, governments committed themselves, *inter alia*, to an open, equitable, rule-based, predictable and non-discriminatory multilateral trading system.

FAO, along with other international organizations, focuses substantial attention on this all-important debate on trade and poverty. FAO, the United Nations Conference on Trade and Development (UNCTAD) and the United Nations Development Programme (UNDP), along with the WTO and the World Bank, have each published recent reports addressing the links between development and trade (see Box 1).

The State of Food and Agriculture 2005 highlights the common lessons, insights and issues – both resolved and unresolved – presented in these and related publications. The report focuses more directly on how trade and poverty linkages can be best used to enhance food security, address inequality and improve overall economic growth.

---

[1] Adopted in September 2000, at the United Nations Millennium Summit, where world leaders also agreed to a set of timebound and measurable goals and targets for combating poverty, hunger, disease, illiteracy, environmental degradation and discrimination against women. These are referred to as the Millennium Development Goals.

Trade, poverty and food security: what are the linkages?

The economic linkages among trade, poverty and food security are complex and national experiences with trade reform have been highly variable. Simple, unambiguous messages are thus difficult to identify, although some policy conclusions can be drawn.

Trade–poverty–food security linkages are diverse in nature. The first linkage occurs at the border. When a country liberalizes its own trade policies by lowering tariffs, for example, this will result in lower market prices for imports at the country's border. When other countries liberalize their trade policies, this will affect border prices of the first country's imports and exports.

The second linkage focuses on how prices are transmitted from the border to local markets within the country: to producers, consumers and households in general. The extent to which households and businesses in the economy experience these price changes depends on the quality of infrastructure and the behaviour of domestic marketing margins, as well as geographical factors.

The empirical literature suggests that the degree of price transmission from the border to the local market can vary widely, even within a single country.

The initial impact of trade liberalization on households occurs once the local market price changes have been determined. Not surprisingly, households that are net sellers of products whose prices rise, in relative terms, benefit in this first round. Net purchasers of such goods lose.

However, the literature also demonstrates that first-round effects are altered significantly in the wake of subsequent household adjustments in consumption and production. In response to changing relative prices, households modify their consumption basket, adjust their working hours and possibly change their occupation. Changes in relative prices can even affect a household's long-term investment in human capital.

As households change their spending levels and employment patterns and as landowners and firms adjust their hiring, a wide range of effects ripple throughout the economy. For example, trade reforms that stimulate agricultural production often lead to a general increase in wages for unskilled labour. This, in turn, benefits households that are net suppliers of unskilled labour.

Finally, the long-run growth effects associated with trade liberalization need to be considered, including increases in firm productivity due to access to new inputs and technologies as well as potential gains due to the disciplining effect of foreign competition on domestic mark-ups.

Exactly how trade affects poverty and food security depends upon each country's specific circumstances, including the situation and location of the poor and food-insecure and the specific reforms undertaken. Understanding and managing these relationships requires country-specific research and country-specific policies. One size does not fit all.

FAO's ongoing studies and analyses, to date, provide encouraging lessons and overall policy guidance. Among the many important lessons is the need for policy-makers to consider more carefully than they have in the past how trade policies can be used positively to promote pro-poor growth. This involves actively implementing policies and making investments that complement trade reforms and enable the poor to take advantage of trade-related opportunities, while establishing safety nets to protect vulnerable members of society.

The analysis presented in this report concludes that multilateral trade liberalization offers opportunities for the poor and food-insecure by acting as a catalyst for change and by promoting conditions in which the food-insecure are able to raise their incomes and live longer, healthier and more productive lives.

It also demonstrates that trade liberalization can have adverse effects, especially in the short run, as productive sectors and labour markets adjust. Opening national agricultural markets to international competition – especially from subsidized competitors – before basic market institutions and infrastructure are in place can undermine the agriculture sector, with long-term negative consequences for poverty and food security. Some households may lose, even in the long run.

To minimize the adverse effects and to take better advantage of emerging opportunities, governments need answers to a wide range of questions: How does trade policy fit into the national strategy to promote poverty reduction and food security? How will the trade reform process and the broader set of economy-wide and sector-specific policies affect relative prices at the border? How will local markets and distribution networks pass on these price effects? What are the expected effects on employment? Which sectors, which parts of the country, and what types of skills will be affected? What will be the revenue effects for the domestic treasury?

Not only are answers to these questions needed, but immediate actions are required. Consistent and sustained policy interventions are needed to provide appropriate signals for pro-poor, pro-growth outcomes to trade. Investments are needed in rural infrastructure, human capital and other public goods. Policy-makers need to give priority to the expenditures and investments that are most essential to the poor and to the long-run viability of their livelihoods.

Safety nets are needed both to protect vulnerable groups from trade-related shocks and to allow the poor to take advantage of economic opportunities arising from trade. Of course, trade and trade reforms are not the only source of shocks faced by the poor and food-insecure. A host of other shocks – natural, human-induced and market-related – can spell disaster in the absence of effective safety nets.

Safety nets are not, however, a substitute for addressing weak institutions, inadequate infrastructure and distorted factor markets, or for making essential investments in health, sanitation and education for rural people. Safety nets merely complement these fundamental actions in preparation for more open markets.

Overview of the report

Chapter 2 presents an overview of patterns of production and trade in agriculture, with particular emphasis on developing countries.

Developing countries are increasingly net importers of food and many have negative net agricultural trade balances. This trend is likely to continue for many developing countries (even if OECD countries reduce their agricultural protection and support policies).

Agricultural exports account for less than 10 percent of the total exports from developing countries, and less than 20 percent in the case of LDCs. Some countries remain much more dependent on agricultural commodity exports, however; these countries are particularly vulnerable to commodity price shocks and weather-related risks.

Over the past two decades, the share of LDCs and net food-importing developing countries (NFIDCs) in global agricultural exports has declined and their share in global food imports has increased. FAO projections suggest a continuing rising trend in the net food imports of developing countries to the year 2030.

The LDCs have seen a rise in their food import bills relative to total export revenues, creating balance of payment difficulties for many of these countries. Many LDCs depend primarily on agriculture for their economic development, so unless they raise their competitiveness in agriculture or diversify their economies, they will become increasingly dependent on aid and more indebted. From a food-security perspective, these countries are particularly vulnerable.

Exports of processed agricultural products are expanding significantly more quickly than those of semi-processed and bulk commodities, and now account for one-half of global agricultural trade. Processed goods offer more possibilities for product differentiation and more opportunities for adding value. They also have a larger potential for intra-industry trade, (i.e. trade that occurs when a country exports and imports goods in the same industry). For example, cocoa-exporting countries are unlikely to import cocoa beans. Chocolate bars, however, are more likely to be exported and imported by the same country. A variety of technical, institutional and market barriers restrict the participation of many developing countries in this more labour-intensive, value-adding growth area.

The share of agricultural trade among developing countries has increased sharply during the past decade, partly as a result of the emergence of regional trade agreements and partly because developing countries represent the key growth markets for agricultural goods. Income growth, urbanization and expanding numbers of women in the labour force are creating new opportunities for increased trade among developing countries, especially in processed food.

Large transnational food companies and supermarkets are influencing domestic food-supply chains through retail procurement logistics, inventory management and distribution networks, and a rapid rise in private standards and gradual rise in the use of contracts.

Chapter 3 examines the trade–policy landscape for agriculture, including an assessment of the reforms that have occurred under structural adjustment programmes and the WTO's Uruguay Round Agreement on Agriculture (AoA).

It is argued that the reform process begun under the Uruguay Round of trade negotiations was an important first step, but has resulted in little real reform of agricultural policies so far. Much remains to be done to complete the multilateral reform process.

Although countries have generally complied with their commitments under the Agreement, international agricultural trade continues to be highly distorted. A review of the state of agricultural protection in the world suggests that protection remains high in many countries, with the highest protection being applied by developed countries and higher-income developing countries. Tariff peaks and tariff escalation create severe distortions that systematically work against the efforts of producers in developing countries to enter the rapidly growing markets for processed products.

Many countries complied with their AoA commitments on domestic support by adopting policy measures that are exempt from disciplines. The degree to which the support measures that are currently exempt are decoupled from production continues to be debated, but the evidence suggests that they are not entirely production-neutral. Further effective disciplines are needed to ensure that domestic support measures are minimally trade-distorting.

Export competition appears to be an area where significant reform is likely in the current Doha Round of trade negotiations. WTO members appear to be ready to eliminate direct export subsidies, although issues of timing and of equivalence with other export competition measures remain contentious. Effective disciplines are needed, but particular care must be exercised to ensure that further disciplines on food aid do not interfere with its humanitarian role.

Developing country experiences of market reforms under structural adjustment have been highly variable: some countries have fully and consistently implemented reforms while others reformed in name only or reversed course unpredictably. Such stop-and-go policies can negate the potential benefits of agricultural and trade policy reforms.

The three so-called "pillars" of the AoA (domestic support, export competition and market access) are interlinked. Many developing countries continue to resist reducing their tariffs as long as their farmers have to compete with subsidized production from other countries.

Chapter 4 surveys some of the most recent economic-modelling exercises that explore the potential economic gains at the national level resulting from serious reforms of the trade and agricultural policies of both developed and developing countries.

Policy-makers need good analytical results in order to understand the potential impacts of alternative policy choices and to devise appropriate measures to ensure that the most vulnerable groups are supported during the trade reform process. The quantitative studies discussed use a variety of modelling approaches and differ significantly in their details. Despite these differences, however, a few consistent conclusions, summarized in the paragraphs below, can be observed.

Agricultural trade reforms could produce important welfare gains at the global level for most, but not all, individual countries. Several recent studies suggest that the largest gains would be achieved under a comprehensive liberalization programme that addresses all economic sectors and all regions. Scenarios in which a single sector or group of countries liberalize would produce far smaller gains.

Industrial countries have the most to gain from agricultural trade liberalization, in absolute terms, because their agriculture sectors are the most distorted by existing policies. Consumers in currently protected markets and producers in countries with low levels of domestic support would tend to gain the most.

The potential gains from agricultural trade liberalization for developing countries, although smaller in absolute terms, would be larger relative to gross domestic product (GDP) because agriculture constitutes a comparatively large share of their economies.

While developing countries as a group stand to benefit from liberalization, some groups could be hurt, at least in the short run. NFIDCs and recipients of preferential access to highly protected OECD markets are vulnerable in this regard.

The liberalization of domestic supports and export subsidies in the OECD countries could result in higher food prices. While producers would benefit from higher commodity prices, consumers would pay higher prices for food. For net food importers, the negative impact on consumers could outweigh the potential benefit to their producers.

Furthermore, developing countries that currently rely on preferential access to OECD countries for their exports could be harmed by reforms that reduce the value of these preferences, unless compensatory measures are put in place.

The net result for these vulnerable countries depends crucially on the policy response of the country itself and the ability of its people to adjust to the changing economic circumstances. This argues for a concerted programme of technical assistance and support for these countries before and during the reform process.

Some developing country exporters would gain as a result of OECD liberalization, but benefits for developing countries are also expected to come from the liberalization of trade among themselves. Indeed, between 70 and 85 percent of the potential benefits for developing countries would result from their own reform policies in agriculture.

Job creation and wage growth for the rural and urban poor constitute one of the main avenues through which trade liberalization can benefit developing countries. Moreover, a broad-based multilateral trade liberalization programme is more likely to benefit the poor than would reforms that focus solely on agriculture and solely on OECD countries. Special attention should be given to labour markets to ensure that the poor are able to make good use of what may be their main asset – their labour.

Chapter 5 takes the analysis from the macroeconomic level to the household level to examine the impact of agricultural trade on poverty.

The results confirm that the primary endowment of the poor is their labour, and that the impact of trade policy reforms on unskilled wages is central to the poverty story, underscoring the importance of domestic policy reforms aimed at improving the functioning of labour markets.

For many developing countries, the principal way in which trade generates positive impacts on poverty and food security is through non-agricultural incomes. Job creation and higher wages in non-agriculture sectors are the biggest promises of trade reform.

Poverty and hunger are also influenced by price changes arising from trade liberalization. The model-based studies discussed in Chapter 4 suggest that net purchasers of agricultural commodities (most of the poor) would be hurt by the higher prices predicted in the wake of comprehensive trade reform.

Higher commodity prices may indeed hurt the poor in the short run, but, in the longer run even net purchasers can benefit if higher commodity prices translate into more jobs and higher wages. The cases reviewed in Chapter 5 suggest that this is often the case. Safety nets and food distribution schemes can also help ensure that low-income consumers are not penalized by rises in the prices of food imports.

Another avenue through which trade reforms can promote pro-poor growth is by removing tariffs on agricultural inputs (machinery, fertilizers and pesticides) in developing countries. Many developing countries continue to penalize their agriculture sectors with these kinds of taxes. Their removal would improve the terms of trade for agriculture and help producers compete on both domestic and international markets.

The evidence presented in this chapter suggests that the trade–growth linkage can be an important vehicle for poverty reduction. However, its potential in this respect depends crucially on effective investments in infrastructure, institutions, education and health.

Chapter 6 examines the significance of trade reform for food security. Food insecurity and poverty are closely interlinked but distinct phenomena. While food insecurity is often a result of poverty, it is also a leading cause of poverty. Hunger and malnutrition can permanently stunt the developmental capacity of children, making it more difficult for them to grow and learn. Hunger has longer-term economic implications because it reduces people's capacity to work and fight disease.

Agricultural trade and trade policy affect food security in many ways. For many policy-makers, tariffs on basic food commodities represent an ongoing dilemma. The justification for such tariffs is often that they offer protection for domestic producers from imports of subsidized commodities; however, they also raise the cost of food, thus taxing the people who can least afford it. This effect has immediate humanitarian implications, of course, because 852 million people in the world lack the ability to grow or buy enough food for their needs.

Trade's contribution to food security involves aspects other than market access in agriculture. It means better trading conditions for non-agricultural products as well, which improves access by the poor and food-insecure to jobs, income, assets and food.

This chapter presents a recent assessment of 15 country case studies undertaken by FAO, examining country experiences of the effects of trade and economic reforms on food security. Although these experiences were highly variable, some general policy lessons can be identified.

First, a country's pre-existing economic structure and policy environment have a strong influence on the results of policy reforms. The existence and functioning of market institutions are particularly important in this regard. In countries where reforms involved the dismantling of state agricultural institutions, finding mechanisms to encourage and assist the private sector to fill these gaps was vital.

Second, countries that implemented targeted transitional measures to protect and compensate vulnerable population groups were more successful in ensuring positive food-security outcomes. Many countries experienced difficulties in implementing safety net programmes effectively.

In addition to safety nets, complementary policies aimed at improving the productivity and competitiveness of the agriculture sector were also essential to positive food-security outcomes. Creating a policy environment to support productive investments by small farmers made it much more likely that they could respond to price incentives and take advantage of the opportunities offered by reform. Improving rural infrastructure was important in most countries, but it was particularly needed in low-income areas.

In countries with a large proportion of low-income and resource-poor people living in rural areas and dependent on agriculture, reforms aimed at raising productivity, creating non-agricultural employment and facilitating the transition out of agriculture were essential for enhancing food security in the medium-to-long term.

However, because such policies may take some time to yield results, they should be set in motion before enacting trade or agricultural policy reforms that may impinge on low-income, food-insecure households. The sequencing of reforms requires special and ongoing attention.

Chapter 7 outlines a twin-track approach to ensuring that the poor and food-insecure are able to capture the potential benefits of agricultural trade and further trade reform. It asks whether the necessary investments are being made to ensure that the poor and hungry are able to share in the gains from trade. Finally, it draws some overall conclusions to the report.

Trade policy reform can offer opportunities to the poor and food-insecure, but the adjustment process must be managed carefully and adequate protection of the vulnerable and food-insecure must be ensured.

Trade liberalization can be a key component for promoting and sustaining agricultural growth. Expanding markets overseas provide farmers with opportunities to supply richer markets and develop brands and qualities that enable them to increase their returns from sales. Liberalization can also create conditions for faster income growth through better access to ideas, technology, goods, services and capital, and by promoting a more efficient use of resources through specialization and the scope for economies of scale. Such growth can also benefit domestic agriculture.

However, the benefits from trade liberalization do not come automatically. Many developing countries need companion policies and programmes that help increase agricultural productivity and product quality if they are to raise their competitiveness in domestic and international markets.

Examples of companion policies include institutional and market reforms, investments in roads, market information systems and related service industries, and policy measures to promote appropriate technological innovations. Above all, countries need to ensure that vulnerable individuals, households and groups that may be disadvantaged by the initial impacts of trade reforms are identified and cushioned through well-designed measures and safety nets.

These policies are described more fully in FAO's twin-track approach, which focuses on (i) creating opportunities for the hungry to improve their livelihoods and (ii) ensuring access to food for the most needy through safety nets and other direct assistance.

Trade policy reforms, like any other potential shock to an economy, entail adjustment costs and not everyone necessarily benefits. Governments in developed countries and developing countries alike have a responsibility to ensure that the reform process is managed in a way that minimizes the risk to vulnerable groups and maximizes their opportunities to share in the gains.

2. Trends and patterns in international agricultural trade

To help understand trade's role in contributing to food security and poverty reduction, this chapter begins with an overview of the role of trade in the world economy. We build on this overview to explore how trade patterns are shifting, contrasting the differences between developed and developing countries in international agricultural trade.[2]

The global economy, including agriculture, is integrating rapidly through trade. At the same time, the exports of developing countries are becoming increasingly diversified, so that these countries are less dependent on agricultural exports than they were in the past. Moreover, developing countries are rapidly becoming their own best markets for agricultural products.

Exports of processed agricultural products are expanding and now account for almost half of global agricultural trade. This phenomenon is being driven by demographic, social and economic trends that are transforming the agricultural and food markets in developing countries. Supermarkets, for example, are rapidly emerging as a major force in developing countries.

The LDCs face particular challenges in world agricultural markets. They are much less integrated into the world economy than are developing countries as a whole, and this feature is particularly striking for their agriculture sectors. As is the case for developing countries as a group, the LDCs have seen their agricultural exports decline as a share of total exports, but their agricultural imports, mostly food, have not fallen as a share of total imports and they now face a large and rapidly growing trade deficit in agriculture.

---

[2] Agricultural data in this chapter include crops, livestock, forestry and fisheries products in bulk and processed forms.

Agricultural trade and the world economy

The past several decades have witnessed a dramatic increase in the integration of the world economy through trade. Figure 1 illustrates the average annual growth rates of global GDP and global exports of goods and services. Global trade in goods and services is expanding more rapidly than global GDP.

International trade in agricultural products has also expanded more rapidly than global agricultural GDP, although at lower rates than for overall trade in goods and services and overall GDP (Figure 2). Slower growth in agricultural output and trade reflects the declining relative importance of agriculture in the world economy and in world trade.

The result of the more rapid expansion of trade (exports plus imports) relative to output is illustrated in Figure 3. Trade intensity, expressed as a ratio of total trade in goods and services to total GDP, has increased from less than 30 percent three decades ago to almost 50 percent today. This trend has been even more dramatic for agricultural trade (including fisheries and forestry), which has grown from around 60 percent to more than 100 percent during the same period. The high trade intensity of agriculture reflects the complementary nature of agricultural production in different agro-ecological zones and a high level of intra-industry trade in the sector.

Nevertheless, the increasing importance of agricultural trade relative to agricultural output has not prevented agricultural trade from losing its relative importance as a component of international trade. Indeed, while agricultural trade continues to expand, its share in total merchandise trade continues to fall, from close to one-third four decades ago to around 10 percent today, as seen in Figure 4.

FIGURE 1
Growth in global GDP and global trade in goods and services
(In nominal terms)

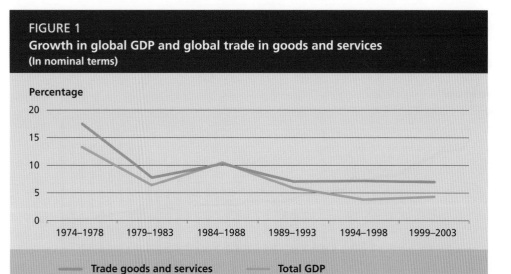

Source: World Bank.

FIGURE 2
Growth in global agricultural GDP and global trade in agricultural goods
(In nominal terms)

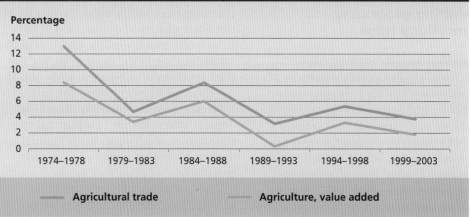

Source: FAO and World Bank.

FIGURE 3
Ratio of trade to GDP in the global economy

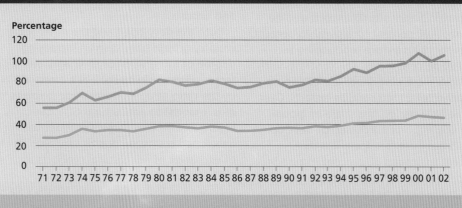

Source: FAO and World Bank.

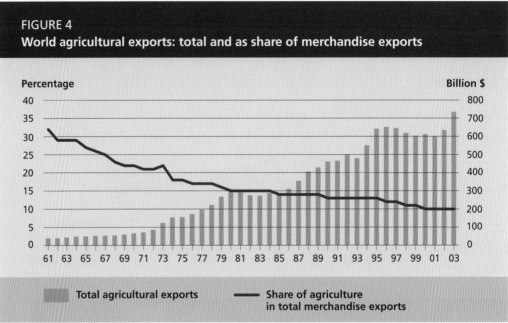

FIGURE 4
World agricultural exports: total and as share of merchandise exports

Source: FAO and World Bank.

The shifting geography of agricultural trade

The past four decades have also seen major changes in geographical patterns of agricultural trade. An increasing share of global agricultural exports originates from developed countries. The European Union (EU) countries account for most of this growth; their share of total agricultural exports has increased from slightly more than 20 percent in the early 1960s to more than 40 percent today. A large portion of this increase is accounted for by intra-EU trade, which represents around 30 percent of world agricultural trade.

Conversely, during the past four decades, the developing countries have seen their share of world agricultural exports decline from almost 40 percent to around 25 percent in the early 1990s before rebounding to around 30 percent today. This contrasts with the steadily increasing share of developing countries in total merchandise exports. Over this same time period, the share of global agricultural imports purchased by developing countries has increased from less than 20 percent to around 30 percent (Figure 5).

The role of agricultural trade in the overall trade patterns has changed in both developed and developing countries. Over the past four decades, the developing countries have seen a major decline in the share of agricultural exports in their total merchandise exports, together with a slower decline in the share of agriculture in their total imports (Figure 6, page 16). They have moved from a positive net agricultural trade position, with exports exceeding imports by a significant proportion, to a situation in which agricultural imports and exports have been roughly balanced in recent years.

Developed countries have seen their share of both agricultural exports and imports decline more slowly over the same period (Figure 6). Today, for both developed and developing countries, agricultural trade is roughly balanced and corresponds to around 10 percent of both total merchandise imports and exports.

Both the developed and the developing country groups have seen an increasing degree of integration of their agriculture sectors into world markets as expressed by the ratio of agricultural trade (exports plus imports) to agricultural GDP (see Figure 7). This is extremely pronounced for developed countries, due to very high levels of exchange of agricultural products in particular among the EU countries.

As seen in Figure 8 on page 18, the role of agricultural trade varies among the

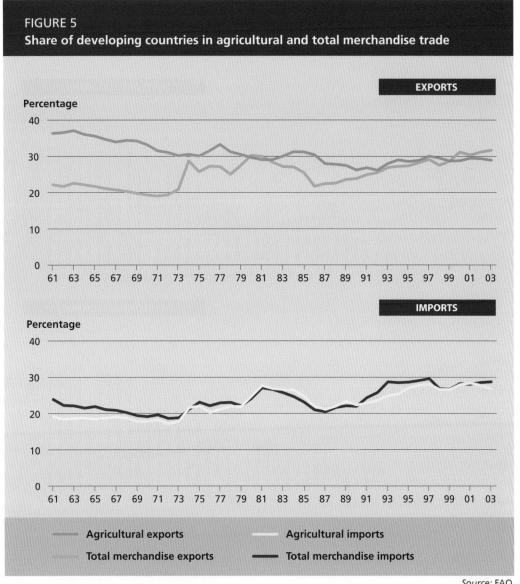

FIGURE 5
Share of developing countries in agricultural and total merchandise trade

Source: FAO.

developing country regions. Only Latin America and the Caribbean region has maintained a strong positive net agricultural exporter position. Indeed, agricultural exports continue to occupy a major share of the region's total merchandise exports, although the share has approximately halved over the past 40 years, from around 50 percent to between 20 and 30 percent in the most recent years.

Sub-Saharan Africa has seen an even sharper decline in the share of agriculture in its exports, from more than 60 percent four decades ago to around 20 percent today. The region remains a net agricultural exporter but with its agricultural imports and exports closer to balance than in the case of Latin America and the Caribbean.

For Asia and the Pacific, both agricultural exports and imports account for less than 10 percent of total exports or imports today; only in the most recent years has the region seen an agricultural net import position.

The Near East and North Africa region is characterized by a significant agricultural trade deficit, which emerged after 1973, as imports expanded rapidly following the oil price boom. Since then, agricultural exports have accounted for at best slightly more than 5 percent of total merchandise exports and agricultural imports now account for around 15 percent of total merchandise imports.

FIGURE 6
Agricultural trade in developed and developing countries

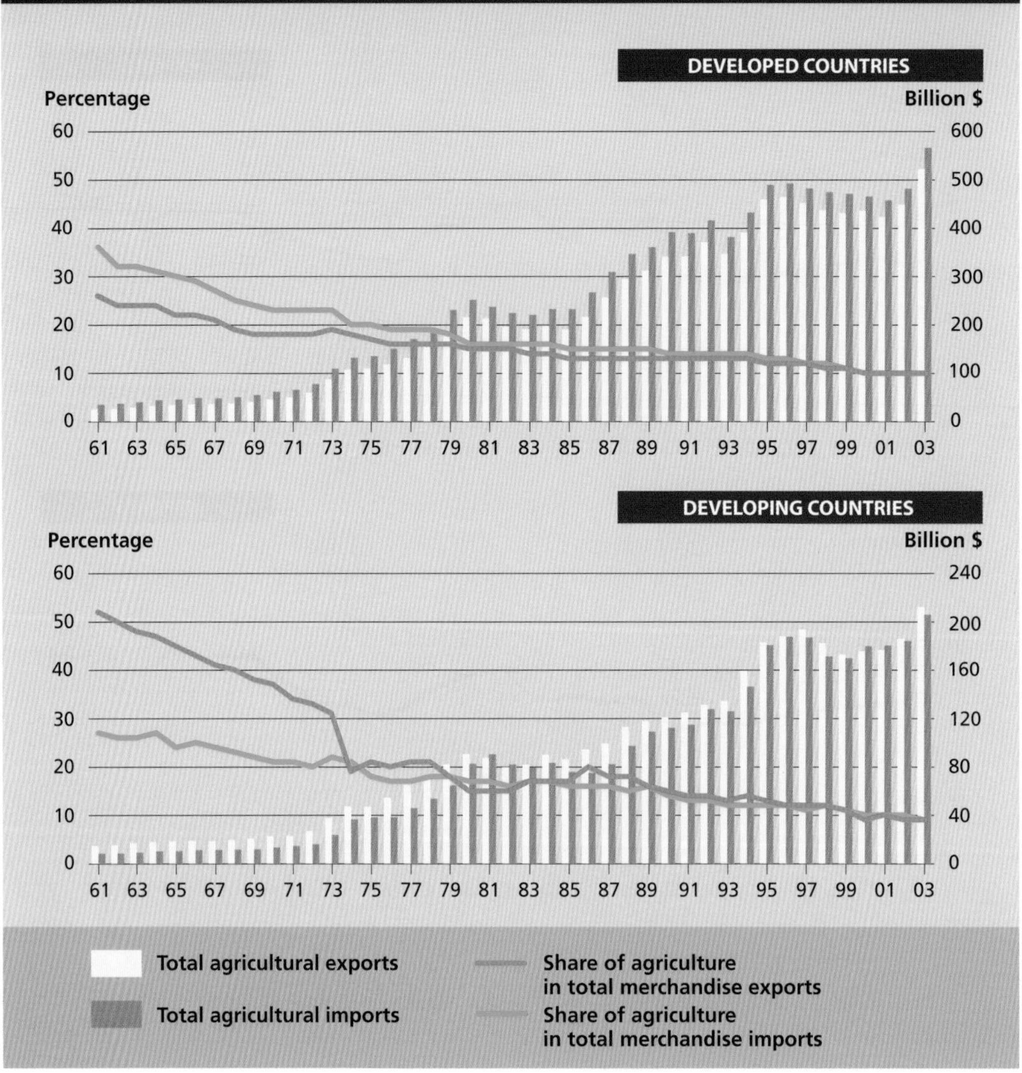

DEVELOPED COUNTRIES

DEVELOPING COUNTRIES

- Total agricultural exports
- Total agricultural imports
- Share of agriculture in total merchandise exports
- Share of agriculture in total merchandise imports

Source: FAO.

The changes in agricultural trading patterns of the developing country regions are reflected also in their share of international agricultural trade (Figure 9, page 19). With the exception of Asia and the Pacific, all regions have seen their share of world agricultural exports decline, although Latin America and the Caribbean region has recaptured some market share in the course of the 1990s.

One of the most striking phenomena evident from Figure 9 is the gradual marginalization of sub-Saharan Africa on international agricultural export markets; the region's share of global agricultural exports has declined gradually from almost 10 percent four decades ago to some 3 percent today. On the import side, the opposite pattern emerges: all developing country regions have seen their share in world agricultural imports increase, with sub-Saharan Africa being the only exception.

Agricultural trade in the least developed countries

The LDCs represent a particular case in terms of long-term trends in global agricultural trade. The agricultural exports of this group

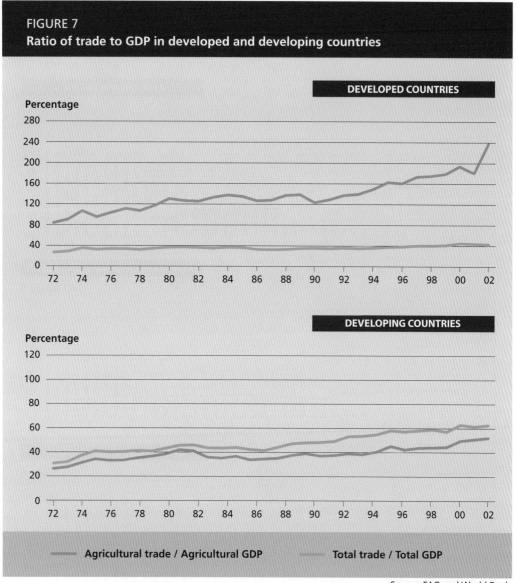

FIGURE 7
Ratio of trade to GDP in developed and developing countries

Source: FAO and World Bank.

of countries have declined dramatically as a share of their overall exports, while agricultural imports have consistently represented around 25 percent of their total imports (see Figure 10, page 20). The LDCs have moved from a position of net agricultural exporters to one of net agricultural importers, and since the late 1980s their agricultural trade deficit has widened rapidly.

At the same time, the LDCs display a strikingly low degree of integration of their agriculture sector into world markets compared with the developing countries overall (Figure 11, page 20, see also Figure 7).

In the mid-1960s, their agricultural trade (exports plus imports) corresponded to slightly more than 20 percent of their agricultural GDP, representing about the same ratio as that of the developing countries overall. Since then, however, the ratio for the LDCs has increased only slightly, to around 30 percent, while for the developing countries overall it has increased to around 50 percent.

Agricultural trade within regions

There has been a tendency in recent decades towards increased intensity of

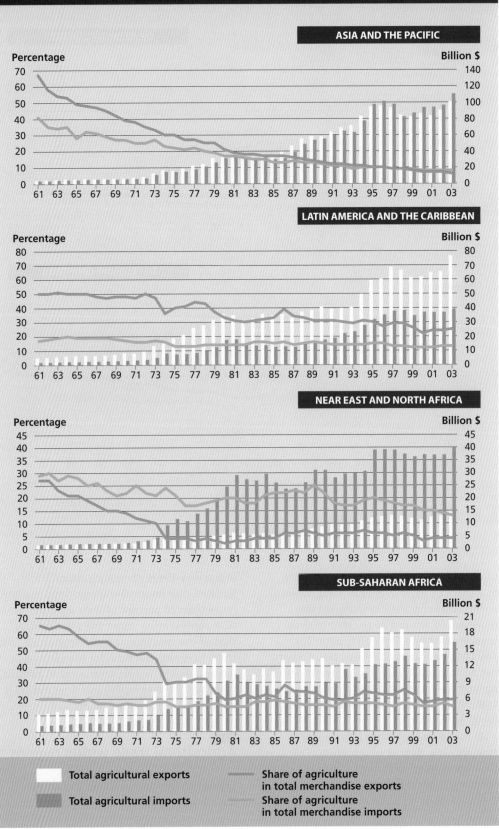

FIGURE 8
Agricultural trade in the developing country regions

Source: FAO.

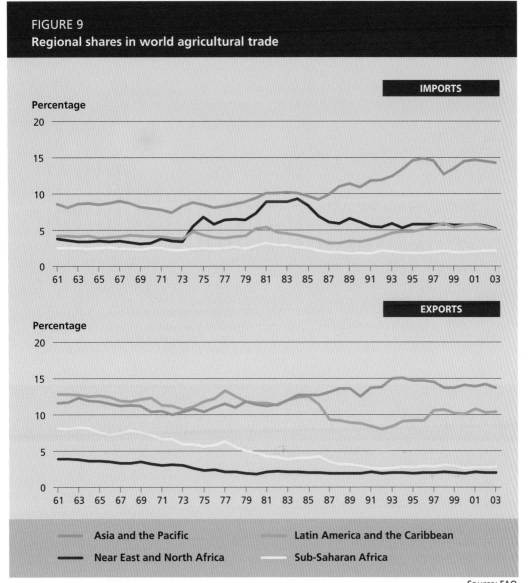

FIGURE 9
Regional shares in world agricultural trade

Source: FAO.

agricultural trade within regions (Tables 1 and 2, pages 22–24).

Within the developed countries, agricultural trade remains largely and increasingly self-centred: some 80 percent of developed country agricultural exports are destined to other developed countries and more than 70 percent of developed country agricultural imports originate in other developed countries.

Particularly significant is the role of trade among EU countries, with more than 70 percent of EU country exports going to, and more than 60 percent of their imports coming from, other EU countries. Agricultural trade among the EU countries represents 30 percent of total world agricultural trade.

Trade between Canada and the United States of America, although much smaller in both absolute and relative terms than intra-EU trade, has grown rapidly since 1980, reflecting the growing importance of the North American Free Trade Agreement (NAFTA), and prior to that the US–Canada Free Trade Agreement, in shaping their trade flows.

In contrast, although agricultural trade among the developing countries has been increasing, particularly during the 1990s, they still depend to a large extent on the developed countries, both as outlets for their agricultural exports and as suppliers of their agricultural imports.

The proportion of developing country agricultural exports going to other

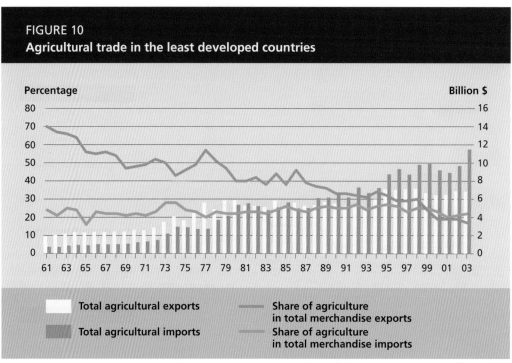

FIGURE 10
Agricultural trade in the least developed countries

Total agricultural exports
Total agricultural imports
Share of agriculture in total merchandise exports
Share of agriculture in total merchandise imports

Source: FAO.

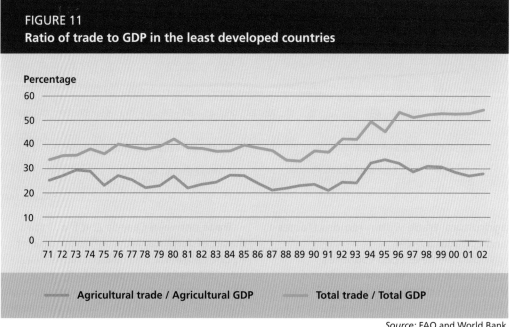

FIGURE 11
Ratio of trade to GDP in the least developed countries

Agricultural trade / Agricultural GDP   Total trade / Total GDP

Source: FAO and World Bank.

developing countries grew from 31 percent in 1990 to 40 percent in 2002, while on the import side the share of developing country imports originating in other developing countries expanded from 36 percent to 45 percent over the same period.

This trend towards increased weight of trade among other developing countries since 1990 is common to all regions and reflects a growing share of agricultural trade taking place within individual regions.

Processed products and the role of supermarkets

The share of processed products in agricultural trade has been increasing for

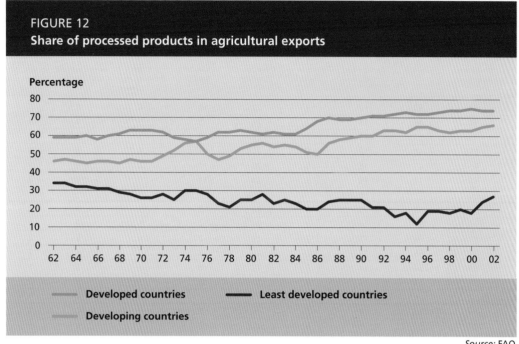

FIGURE 12
Share of processed products in agricultural exports

Source: FAO.

both developed and developing countries, but remains much higher for the former group (see Figure 12).

There are wide differences among the developing countries. For the LDCs, the share of processed products in total agricultural trade is significantly lower than for the overall group and has been gradually declining. Only in the second half of the 1990s did the share of processed products in LDC agricultural exports begin to increase.

The rapid growth in processed agricultural trade has less to do with agricultural trade policy reform than with the massive demographic and economic transformations that are sweeping through the developing world. Urbanization, the participation of women in paid employment and rising incomes have increased the opportunity cost of purchasing and preparing bulk foods and help explain the rapid shifts towards processed foods that are occurring in both international trade and domestic markets.

Related to the growth in processed food trade is the rapid expansion of supermarkets in many developing countries. Research suggests that in Latin America, where this trend is most advanced, the quantity of fruits and vegetables purchased from local producers by supermarkets to supply local stores is 2.5 times higher than the total exports of produce from Latin America to the rest of the world (Reardon and Berdegué, 2002).

Many of the supermarkets that are emerging in developing countries are owned by multinational chains based in Europe, Japan and North America. These companies face saturated markets and intense competition in their home markets and they have been attracted by the higher profit margins to be obtained by investing in these new markets. The liberalization of policies governing foreign direct investment in the retail sector has facilitated the trend.

These global chains diffuse management practices that promote efficiency in logistics and inventory management, leading to centralized procurement and consolidated distribution patterns. The organization of retail trade is being transformed in terms of larger volume per supplier, fewer suppliers, the rapid disappearance of small family-owned retailers and a reduction in the role of central markets. A shift is occurring away from traditional wholesalers and brokers towards specialized wholesalers and towards export firms with new domestic market functions. Agrifood market institutions are being affected also, with a rapid rise in private standards and a gradual rise in the use of contracts.

TABLE 1
Destination of agricultural exports by region (percent)

Exporting region	Year	Destination								
		Developed countries	EU-15	Canada and United States	Countries in transition	Developing countries	Asia and the Pacific	Latin America and the Caribbean	Near East and North Africa	Sub-Saharan Africa
World	1980	73	50	10	3	27	10	6	8	3
	1985	74	48	15	2	26	8	5	9	3
	1990	78	53	13	2	22	8	5	7	2
	1995	75	48	12	5	25	11	6	6	2
	2000	75	43	15	5	25	11	6	6	2
	2002	76	44	15	6	24	11	6	6	2
Developed countries	1980	74	53	9	3	26	8	7	9	3
	1985	76	53	14	2	24	7	5	9	3
	1990	81	60	11	2	19	7	4	6	2
	1995	80	55	11	6	20	8	5	5	2
	2000	81	50	14	6	19	8	5	5	1
	2002	81	51	14	6	19	8	5	5	2
EU-15	1980	82	76	4	3	18	2	3	10	4
	1985	84	76	8	2	16	2	2	9	3
	1990	88	82	5	2	12	2	2	6	2
	1995	89	79	4	6	11	2	2	5	2
	2000	89	73	6	5	11	3	2	5	2
	2002	90	73	6	6	10	2	1	4	2
Canada and United States	1980	65	29	14	3	35	13	14	6	2
	1985	67	22	22	1	33	11	12	7	3
	1990	70	22	24	1	30	12	10	6	1
	1995	67	18	25	2	33	15	11	6	1
	2000	67	13	32	1	33	12	14	6	1
	2002	65	12	35	1	35	13	15	5	1
Countries in transition	1980	89	63	8	14	11	1	1	8	1
	1985	88	58	7	18	12	2	2	8	0
	1990	88	66	4	12	12	1	1	10	1
	1995	92	44	1	46	8	2	0	5	0
	2000	87	39	2	39	13	5	0	7	0
	2002	84	40	2	36	16	8	0	8	0
Developing countries	1980	67	35	14	2	33	18	5	7	2
	1985	69	35	17	3	31	13	5	11	2
	1990	69	33	19	2	31	14	6	8	2
	1995	63	28	17	3	37	19	8	8	3
	2000	62	25	19	3	38	18	8	8	3
	2002	60	25	18	4	40	19	8	9	4
Asia and the Pacific	1980	61	24	12	2	39	27	2	9	2
	1985	62	21	14	3	38	24	1	10	2
	1990	63	21	12	2	37	26	2	8	2
	1995	57	16	12	3	43	32	1	7	2
	2000	57	15	14	2	43	32	1	7	2
	2002	52	14	12	3	48	35	1	8	3

TABLE 1 (cont.)

Exporting region	Year	Destination								
		Developed countries	EU-15	Canada and United States	Countries in transition	Developing countries	Asia and the Pacific	Latin America and the Caribbean	Near East and North Africa	Sub-Saharan Africa
Latin America and the Caribbean	1980	75	44	23	4	25	2	18	4	1
	1985	75	43	25	3	25	4	11	9	2
	1990	75	39	31	2	25	4	14	6	1
	1995	67	33	27	3	33	8	19	5	2
	2000	68	28	30	3	32	7	18	5	1
	2002	66	28	28	4	34	10	17	7	2
Near East and North Africa	1980	72	65	1	5	28	2	0	23	4
	1985	61	44	7	4	39	2	0	35	1
	1990	68	51	6	4	32	3	1	27	1
	1995	64	43	5	11	36	4	1	30	1
	2000	63	42	5	10	37	4	1	30	2
	2002	58	38	4	10	42	6	1	33	2
Sub-Saharan Africa	1980	85	71	9	2	15	3	0	3	8
	1985	86	71	14	2	14	4	0	2	8
	1990	75	67	5	1	25	5	0	2	18
	1995	71	59	5	2	29	8	1	5	14
	2000	61	46	5	2	39	11	1	8	19
	2002	66	50	6	3	34	8	1	6	20

Source: UN Comtrade database.

TABLE 2
Origin of agricultural imports by region (percent)

Importing region	Year	Origin								
		Developed countries	EU-15	Canada and United States	Countries in transition	Developing countries	Asia and the Pacific	Latin America and the Caribbean	Near East and North Africa	Sub-Saharan Africa
World	1980	67	36	23	2	33	12	13	2	6
	1985	66	37	21	2	34	13	14	2	5
	1990	71	43	21	2	29	11	11	2	4
	1995	69	39	22	4	31	13	12	2	4
	2000	69	36	20	5	31	13	13	2	3
	2002	69	38	18	5	31	12	13	2	3
Developed countries	1980	68	38	23	2	32	10	13	2	7
	1985	67	41	20	2	33	11	15	2	6
	1990	73	48	19	2	27	10	12	2	4
	1995	72	46	19	4	28	11	12	2	4
	2000	72	43	18	5	28	11	12	2	3
	2002	73	46	16	5	27	10	12	2	3
EU-15	1980	73	54	16	3	27	6	10	2	8
	1985	73	59	10	3	27	6	12	2	7
	1990	79	68	8	3	21	5	9	2	5
	1995	79	68	8	4	21	5	9	2	5
	2000	79	63	7	4	21	5	9	2	4
	2002	80	65	5	5	20	5	9	2	4

TABLE 2 (cont.)

Importing region	Year	Origin								
		Developed countries	EU-15	Canada and United States	Countries in transition	Developing countries	Asia and the Pacific	Latin America and the Caribbean	Near East and North Africa	Sub-Saharan Africa
Canada and United States	1980	54	14	29	1	46	10	30	1	5
	1985	57	19	30	1	43	10	28	1	4
	1990	60	17	34	1	40	11	27	1	2
	1995	59	14	38	1	41	13	25	1	2
	2000	61	15	38	1	39	13	24	1	1
	2002	61	16	37	1	39	13	24	1	1
Countries in transition	1980	68	31	22	8	32	6	19	4	2
	1985	64	28	6	20	36	13	17	5	1
	1990	67	43	7	12	33	6	13	8	7
	1995	80	44	5	29	20	5	9	2	3
	2000	77	35	6	33	23	7	10	3	2
	2002	74	38	6	27	26	8	13	3	3
Developing countries	1980	62	24	26	2	38	23	9	2	4
	1985	64	23	27	2	36	21	10	3	3
	1990	64	22	30	1	36	19	11	3	3
	1995	59	18	30	3	41	20	14	3	3
	2000	58	16	28	3	42	20	15	4	4
	2002	55	12	27	5	45	21	17	4	4
Asia and the Pacific	1980	55	7	31	0	45	38	5	1	2
	1985	58	9	30	1	42	34	5	1	2
	1990	59	10	32	0	41	31	6	1	3
	1995	57	10	31	3	43	31	7	1	3
	2000	55	10	24	4	45	32	8	1	4
	2002	52	8	22	5	48	34	10	1	3
Latin America and the Caribbean	1980	66	22	40	0	34	3	31	0	0
	1985	65	17	45	0	35	3	31	0	0
	1990	67	18	45	1	33	3	29	0	0
	1995	59	14	42	1	41	3	37	0	1
	2000	61	10	48	0	39	3	35	0	1
	2002	63	8	51	0	37	3	34	0	1
Near East and North Africa	1980	69	43	16	5	31	10	8	7	7
	1985	70	42	17	4	30	12	7	8	3
	1990	69	41	19	4	31	11	7	10	3
	1995	67	34	22	8	33	11	8	10	3
	2000	62	29	19	6	38	13	10	12	4
	2002	57	23	16	10	43	10	15	14	4
Sub-Saharan Africa	1980	70	48	17	0	30	12	5	1	12
	1985	67	47	14	0	33	12	12	1	7
	1990	62	53	7	0	38	16	2	2	18
	1995	54	40	12	1	46	21	9	2	15
	2000	51	33	11	1	49	16	7	2	23
	2002	45	27	11	1	55	22	8	3	22

Source: UN Comtrade database.

These emerging supermarkets do not cater only to higher-income consumers. In Asia and Latin America, they are expanding into poorer neighbourhoods, smaller towns and rural areas, targeting lower- and middle-class consumers. The purchasing practices of the supermarkets are transforming agrifood markets in developing countries, with important implications for small farmers and rural communities. Fundamental changes in the smallholder farming structure need to occur in order for them to be able to supply according to the supermarkets' standards.

Farmers have to produce what the supermarkets demand in terms of both quantity and quality. They often have to comply with certification requirements that are more stringent than official food safety standards. Meeting the demands of procurement officers requires levels of skill and technology that small farmers often do not have. They may have to invest – individually or collectively – in irrigation, greenhouses, trucks, cooling sheds and packing technologies, among other things. They need to be able to sort and grade their produce, meet timing and delivery requirements, and document their farming practices. In addition, they need to be able to bridge the gap between delivery and payment, which presupposes access to credit.

Downstream, the changes have potential benefits for consumers in terms of greater variety, reduced seasonality and lower prices of food products. In terms of food safety, tracing the source of food contamination has become easier, but there is also a risk that unsafe food that may enter into the food chain is distributed rapidly.

Key findings

Several key trends in international agricultural trade have important implications for small farmers and the food-insecure. While policy-makers are gradually recognizing these implications, local, national and international policy and programmes need to adapt rapidly if they are to avoid further marginalizing these groups.

- The global economy, including agriculture, is integrating rapidly through trade. The rate of growth in agricultural trade over the past decade has been about 3 percent annually, more than three times the rate of agricultural output growth.
- Developing countries are much less dependent on agricultural exports than they were in the past.
- Developing countries are rapidly becoming their own best markets for agricultural exports.
- Exports of processed agricultural products are expanding rapidly, driven by demographic, social and economic trends.
- Supermarkets are emerging as a major force in developing countries.
- The LDCs are now much less integrated into the world economy than are developing countries as a whole, and this is particularly striking for their agriculture sectors.
- In contrast with the growing importance of processed agricultural exports in other developing countries, the LDCs have seen the share of their processed product exports decline from around 30 percent of total agricultural exports in the 1960s to less than 20 percent in the 1990s.

3. The agricultural trade policy landscape

Agricultural products have been traded for millennia as people have sought more stable and diverse sources of food. In turn, governments have intervened in agricultural production and distribution systems almost from the beginning of recorded history. Indeed, ensuring adequate supplies of food was one of the earliest tasks undertaken by governments in societies as diverse as the Roman and Incan empires (Woolf, 2003; Crow, 1992).

Governments have used a variety of policy tools to pursue their food and agriculture policy objectives – ranging from trade taxes and production quotas to import monopolies and export bans. While their policy objectives and tools have varied over time depending, among other issues, on the level of economic development and the role of agriculture in their economies and societies, governments around the world continue to view food and agriculture as an essential policy domain.

Quite often, governments pursue conflicting policy objectives. For example, efforts to support farm incomes through market price supports or import barriers could undermine national food security goals by raising food prices for poor consumers. Trade-offs between such competing objectives are usually made at the national level, with different segments of society vying for their own interests within the political system.

Similarly, at the international level, the policy interventions of one country may conflict with those of another, as when efforts to support producers at home hurt producers in foreign countries. The potential for industrial tariffs and subsidies in one country to harm the interests of another country has long been recognized by the international community, but the same problem in agriculture has been acknowledged and addressed only recently.

This chapter reviews briefly the evolution of agricultural trade policy since the middle of the twentieth century, emphasizing the accomplishments of the WTO's Uruguay Round Agreement on Agriculture (AoA) and the unfinished reform agenda currently being discussed in the Doha Round of trade negotiations. Issues regarding the measurement of agricultural support and protection are discussed and comprehensive data and estimates of the actual levels of subsidies and protection being applied on agriculture by countries around the world are presented.

Evolution of agricultural trade policy

Competing agricultural policies

Before the AoA came into force in 1995, the agriculture sector had been excluded from the disciplines of the multilateral trade system. As a result, no institutional mechanism existed to balance the policy interests of different countries. The General Agreement on Tariffs and Trade (GATT), the precursor of the WTO, came into force in 1947 to regulate international trade. The GATT prohibited the use of quantitative import barriers and most domestic and export subsidies for manufactured products and during successive rounds of negotiation reduced import tariffs on manufactured goods to low levels.

The GATT provided specific exceptions for agricultural products, however, and discussion of agricultural policy was kept largely outside the GATT framework. Over time, agricultural trade policies evolved in ways that differed radically from those applied for manufactured goods, with a host of domestic and export subsidies and non-tariff barriers emerging, including variable levies, minimum import prices, voluntary export restraints and quantitative import quotas.

These policies increasingly became a source of international friction. For example, domestic agricultural subsidies were used by many developed countries to guarantee farmers an "adequate" income. Production subsidies such as minimum market support prices tended to stimulate production far beyond the capacity of the domestic market to absorb, generating surpluses that were purchased and held by governments. Some governments then used export subsidies to sell the resulting surpluses on world markets. The United States and the EU, in particular, found their competing agricultural policies to be increasingly expensive and difficult to sustain.

Developing countries in crisis

From the 1950s to the 1970s, the dominant development paradigm involved a strategy of "import substitution" to promote rapid industrialization. Under this strategy, the agriculture sector was taxed heavily to support industrial development, primarily concentrated in the cities. Explicit taxes on agricultural commodity exports were common, but implicit taxes in the form of overvalued currency exchange rates, high industrial import tariffs and subsidies for industrial production were more pervasive.

The "urban bias" embodied in these explicit and implicit taxes systematically discriminated against the agriculture sector and rural areas (Schiff and Valdés, 1998). Many governments attempted to correct the bias against agriculture by intervening in agricultural output and input markets through price measures, compulsory state monopolies and the provision of basic services to the sector (e.g. credit, essential inputs, technical and market information, and marketing and distribution infrastructure). These interventions were often needed to overcome widespread market failures, but they sometimes created additional distortions and rigidities that hampered the sector (FAO, 2005a).

Some poorer countries also imposed trade measures that hurt their neighbours. For example, import quotas were widely used to help stabilize domestic prices in developing countries, but these measures shifted the burden for balancing domestic supply and demand onto world markets, making prices for farmers and consumers in other countries more volatile.

Although many developing countries experienced periods of relatively rapid economic growth at the macro level under these policies, by the late 1970s and early 1980s unsustainable fiscal and current account deficits, hyperinflation, external debt problems and foreign exchange crises signalled the need for policy reform. With the encouragement and support of the IMF and the World Bank, many countries embarked on structural adjustment programmes.

At the macro level, the principal policy-reform strategy involved import tariff reduction, market deregulation, privatization and fiscal stabilization pursued through currency realignments and significant budget cuts. For agriculture, the primary objective was to make the sector more market-oriented. Specific budget cuts were often made in subsidized credit and inputs and in investments in research and infrastructure.

Agricultural reforms typically involved the replacement of most quantitative import restrictions with tariffs; a reduction in both the level and dispersion of tariffs; the removal of export taxes, quotas and licences; the reduction or elimination of state trading; the elimination of domestic price controls and the gradual removal of state procurement programmes (FAO, 2005a).

Multilateral disciplines on agriculture – the Uruguay Round

Against this background of "disarray", the GATT signatory countries embarked on the Uruguay Round of trade negotiations in 1986. The goal of the agricultural negotiations was:

> ... to establish a fair and market-oriented agricultural trading system ... through ... strengthened and more operationally effective GATT rules and disciplines ... resulting in correcting and preventing restrictions and distortions in world agricultural markets.
>
> (GATT, 1994, p. 39)

The Uruguay Round Agreement on Agriculture, which came into force in 1995, represented the first occasion on which a clear set of rules was set up to cover agricultural trade. Although the Uruguay Round has been sharply criticized for failing to secure a significant reduction in support and protection to agriculture, it has been credited with

> BOX 2
> **Main provisions of the Uruguay Round Agreement on Agriculture**
>
> **Domestic support**
> - *Reduction of domestic support.* Reduction commitments on support to agriculture were expressed in terms of a total aggregate measurement of support (total AMS), which is the sum of expenditures on non-exempted support aggregated across commodities and policies. The Agreement called for a 20 percent reduction in total AMS over five years (13.3 percent over ten years for developing countries and no reduction required for LDCs). The reduction commitments applied to total AMS and were not product- or policy-specific.
> - *Exempt policies.* Policies considered as having no or minimal trade-distorting effects or effects on production were exempted from reduction commitments (and could even be increased) and excluded from the AMS. These so-called "green box" policies must not entail price support to producers and must be provided by publicly financed programmes not involving transfers from consumers. The list of specific exempt policies is very long and includes general services, food security stocks, domestic food aid and certain direct payments to producers. In addition, the so-called "blue box" measures exempted direct payments under production-limiting programmes, provided that certain conditions are met.
> - De minimis *exemption:* This allows any support for a particular product to be excluded from the AMS and the corresponding reduction commitment, provided the support does not exceed 5 percent of the value of the total production for the commodity in question, or 5 percent of the value of total agricultural production for non product-specific support. For developing countries, the *de minimis* ceiling is 10 percent.
>
> **Export competition**
> - *Export subsidies.* The AoA defined export subsidies that were to be reduced: direct subsidies, government sales from stocks at prices below domestic prices, export payments

establishing a framework for the progressive reduction of trade-distorting protection of the agriculture sector. This section outlines some of the implications of the Uruguay Round Agreement and the unfinished agenda that is on the table in the Doha Round.

The AoA established disciplines on agricultural policy in three main categories: domestic support, export competition and market access (see Box 2 and below). The three categories were agreed because they are interrelated and mutually reinforcing.

Doha Development Round: Framework Agreement

The AoA included a commitment to further progressive liberalization of the sector. A new round of negotiations was launched in Doha in November 2001. This round, called the "Doha Development Round", is mandated to accord particular priority to the needs of developing countries. On 31 July 2004, the WTO's 147 Member Governments approved a Framework Agreement (WTO, 2004b) and other agreements aimed at advancing progress and successfully concluding the Doha Development Round of trade negotiations. Annex A of the document specifically provides the framework for establishing modalities in agriculture.

The Framework Agreement affirms that:
> Agriculture is of critical importance to the economic development of developing country Members and they must be able to pursue agricultural policies that are supportive of their development goals, poverty reduction strategies, food security and livelihood concerns.
> (para. 2)

Furthermore:
> Having regard to their rural development, food security and/or livelihood security needs, special and differential treatment for developing countries will be an integral part of all elements of the negotiation ...
> (para. 39)

financed by obligatory levies, subsidized export marketing costs and special domestic transport charges. The volume of subsidized exports was to be reduced by 21 percent and the expenditure on export subsidies by 36 percent over five years (for developing countries by 14 and 24 percent, respectively, over ten years). Reductions were to be product-specific. Countries not using export subsidies during 1986–90 were prohibited from introducing them.

Market access
- *Tariffication.* Non-tariff barriers (quotas, variable levies, minimum import prices, discretionary licensing, state trading measures, voluntary export restraint agreements and similar border measures) were abolished and converted to an equivalent tariff, either specific or *ad valorem*. Developing countries were given the option of introducing bound tariff ceilings rather than calculated tariff equivalents.
- *Tariff reduction.* Tariffs, including those resulting from tariffication, were reduced by 36 percent on average over six years, starting in 1995, with a minimum reduction of 15 percent for each item (for developing countries the equivalent reductions were 24 and 10 percent, respectively; LDCs were exempt from reduction commitments).
- *Minimum access.* Where there were no significant imports, minimum access for quantities of imports corresponding to around 3 percent (rising to 5 percent) of domestic consumption in 1995 were to be ensured. Minimum access opportunities were to be implemented through tariff rate quotas (see Box 3).
- *Current access guarantee.* Current access (i.e. the quantity of imports in the 1986–88 period) was to be guaranteed in the event that it exceeded the minimum access level mentioned above.
- *Special safeguard provisions.* These allowed additional duties in the case of import surges (defined by specified trigger levels) or particularly low prices (both compared with 1986 levels).

The document refers to special and differential treatment in the areas of domestic support, export competition and market access to benefit developing countries. There is a commitment to the identification of "sensitive products" and "special products", which will be eligible for more flexible treatment and to a "special safeguard mechanism" for developing countries.

The Framework Agreement provides some flexibility for developed countries but reaffirms their commitment to reform. With reference to the Doha Ministerial Declaration, which calls for "substantial reductions in trade-distorting domestic support", the Agreement states that "there will be a strong element of harmonisation in the reductions made by developed Members. Specifically, higher levels of permitted trade-distorting support will be subject to deeper cuts." A timeline for the elimination of export subsidies is to be established and as a guiding principle for further negotiations on market access the Agreement indicates that "substantial overall tariff reductions will be achieved as a final result from negotiations".

This is to the advantage of both developed and developing countries that have an interest in penetrating export markets. In the areas of market access and domestic support, a tiered formula is called for that represents a single approach for developed and developing country Members and at the same time recognizes their different tariff structures and levels of domestic support.

The sections below examine the existing disciplines under what are referred to as the "three pillars" of the AoA – domestic support, export competition and market access – and assess the progress made thus far in reducing trade-distorting support and protection to the sector. Particular challenges in the ongoing negotiations are highlighted.

Domestic support[3]

The AoA includes disciplines on domestic support in recognition of the potential of such policies to distort production and trade. All domestic support programmes defined as having distorting effects on trade or production were included in the Aggregate Measurement of Support (AMS) and countries agreed to reduce the AMS during the implementation period. Policies defined as having "no, or at most minimal, trade-distorting effects or effects on production" were categorized as "green box" measures and were exempt from the reduction commitments.

Further exemptions were granted for certain direct payments under production-limiting programmes and for supports below a *de minimis* level. Most developing countries declared their domestic agricultural support programmes under the *de minimis* category, although a few reported development-oriented expenditures that are specifically exempt under the provisions for special and differential treatment for developing countries.

Measuring domestic support to agriculture

Different indicators have been developed as measures of support to producers. The two most widely cited are the WTO's AMS and the OECD's producer support estimate (PSE). Although the two indicators take a broadly similar approach, there are a number of methodological differences, and they were developed for different purposes. The AMS is the basis for a legal commitment to reduce domestic support in the WTO AoA, whereas the purpose of the PSE is to monitor and evaluate progress in agricultural policy reform.

The main components of AMS are (i) market price support as measured by the gap between a fixed world reference price and the domestic administered price (which may not be the same as the current domestic market price) and (ii) the level of budgetary expenditures on domestic support policies that are considered to be trade-distorting.

The OECD's PSE indicates the annual monetary transfers to farmers from policy measures that (i) maintain a difference between domestic prices and prices at the country's border (market price support) and (ii) provide payments to farmers, based on criteria such as the quantity of a commodity produced, the amount of inputs used, the number of animals kept, the area farmed, or the revenue or income received by farmers.

Like the AMS, the PSE includes a price gap as well as the level of budgetary expenditures by governments, but there are two key distinctions:

- The market price support in the PSE is measured at the farmgate level using actual producer and border prices for commodities in a given year, whereas in the AMS it is calculated using the difference between the domestic administered support price and a world reference price fixed in terms of a historical base period (1986–88).
- The PSE covers all transfers to farmers from agricultural policies, whereas the AMS covers only domestic policies in the "amber box" category and excludes production-limiting policies ("blue box"), policies that are minimally trade-distorting ("green box") and a *de minimis* level of trade-distorting policies. The result is that trends in the two indicators

TABLE 3
OECD producer support estimate

	1986–88	2001–03	2001	2002	2003[1]
All OECD countries: Value *(million $)*	241 077	238 310	227 955	229 691	257 285
Percentage	37	31	31	31	32

[1] Provisional.
Source: OECD, 2005.

---

[3] The material in this section draws on FAO (2005b).

TABLE 4
Measures of domestic support

	OECD measures		WTO measures of domestic support[1]						
	PSE	PSE minus border protection	Exempt			AMS			
			Green box	Blue box	De minimis	Ceiling	Notified		
	(Million $)		(Million $)			(Million $)	(Million $)	(Percentage of ceiling)	(Percentage consumer-financed)
EU	115 470	75 333	21 261	21 114	18.6	74 102	51 084	68.9	95.0
United States	54 433	21 597	30 591[2]	–	29.1	19 899	16 862	84.7	35.1
Japan	53 991	49 070	23 664	817	91.7	36 461	6 588	18.1	82.1
Republic of Korea	18 308	17 555	4591	–	68.7	1 578	1 306	82.8	100.5
Mexico	4 166	2 666	575	–	–	3 614	500	13.8	91.0
Canada	3 709	2 094	1 177	–	114.0	3 016	632	21.0	46.8

[1] Most recent available data.
[2] The United States has an additional $33 050 million in the green box for domestic food aid.
Source: FAO, 2005b, based on de Gorter (2004), from OECD and country notifications to the WTO.

since 1986–88 show a marked difference. While the AMS has fallen significantly, the PSE has remained relatively stable. Table 3 summarizes the PSE for all OECD countries since 1986–88. While the PSE has fallen as a percentage of the value of agricultural production in the OECD countries, in monetary terms the PSE was higher in 2003 than in the base period. In contrast, the AMS for all WTO members has fallen from over $160 billion to about $60 billion (FAO, 2005b) over the same period.

Table 4 compares the 2003 PSE figures for selected WTO members with their levels of domestic support as measured under the AoA. The first column reports the PSE whereas the second column subtracts the component of the PSE that is provided by border protection, yielding a measure that more closely approximates domestic support to agriculture. The WTO measures are divided into exempt ("green box", "blue box" and *de minimis*) and non-exempt, or AMS, categories. Under the AMS, the ceiling represents the maximum amount of support the country is permitted to provide under its AMS commitments. The figures notified represent the actual amount of AMS expenditures reported to the WTO. For all countries in the table, notified AMS expenditures were below the permitted ceilings. The final two columns of the table show the notified AMS as a percentage of the ceiling and the share of the AMS that is provided by consumers through market prices rather than through transfers from taxpayers.

The vast majority of AMS expenditures are accounted for by the EU, Japan and the United States, with several other OECD countries reporting relatively high AMS levels. Most OECD countries were able to meet their AMS reduction commitments by reformulating their policies to satisfy the criteria for "green box" or "blue box" exemptions. Furthermore, since the AMS commitments are not commodity-specific, some countries met their commitments by reallocating expenditures among commodities within the AMS (Tangermann, 1998). Thus, although the countries having AMS commitments are generally agreed to have met the requirements of the AoA, and some policies have been redesigned to be less trade-distorting, the overall level of support to agriculture in these countries (measured by economic criteria rather than the negotiated criteria used in the Agreement) has fallen very little, if at all.

An unfinished agenda on domestic support

A major criticism of the domestic support provisions of the AoA is that they are unbalanced in the treatment of developed and developing countries. Because most developing countries did not declare domestic support under the AMS, they are constrained to provide support only under the *de minimis* provisions or other exempt policies. It is argued that developing countries lack the administrative or budgetary capacity to implement most "green box" policies, for example, and thus should be allowed to use policies such as domestic price supports that would be categorized under the AMS.

This criticism is weakened because most developing countries are currently providing far less support than is permitted under the *de minimis* provisions, which for developing countries are 10 percent per commodity and 10 percent of the total value of agricultural production. Of more serious concern are the continued high levels of support and protection in some developed countries and whether developing countries should be permitted to provide offsetting protection for their farmers. This topic is explored more fully in the section on market access below.

A more fundamental criticism of the AoA concerns the degree to which different types of domestic support measures are in fact decoupled from production and trade. Empirical evidence on the degree to which exempt domestic supports (as defined by the WTO) distort production and trade is limited because they have only been in operation for a relatively short time – since the 1992 Common Agricultural Policy (CAP) reforms in the EU and the 1996 Farm Bill in the United States. The OECD has conducted simulation exercises to predict the production-distorting effects of alternative domestic support payments relative to the equivalent amount of direct market price support (Anton, 2004). The results suggest that direct payments based on the area planted to a single crop are only 36 percent as production-distorting as market price supports. If the direct payments are further decoupled (i.e. made on total area planted regardless of the crop) their distortiveness falls to less than 20 percent of the distortion caused by market price supports.

Decoupled support to agriculture could influence production decisions through a number of mechanisms beyond the subsidy effects described above. Direct payments influence farmers' perception of risk by changing their wealth status and by providing a form of insurance. They may also influence farmers' decisions about whether to continue farming or exit from the sector. Other factors related to policy design, costs of compliance and enforcement, programme size and the combinations of policies can also influence production decisions.

Several studies have attempted to measure the significance of these so-called "non-price effects". Although partial in their coverage, most of these studies reach a general consensus that non-price effects can be more significant than the subsidy effects reported by Anton (2004). Research from the OECD (2004) suggests that commodity-specific area payments serve to reduce the risk associated with crop production, and that incorporating this insurance effect increases the degree of production distortion associated with these payments to 45 percent of that provided by an equivalent level of market price support. Young and Westcott (2000) argue that crop-insurance schemes that are not commodity-specific implicitly provide different subsidies to individual commodities depending on their relative net returns, with riskier commodities receiving a higher implicit subsidy.

Considerable debate surrounds the impact of decoupled payments on the level and quality of resources devoted to agricultural production. Depending on the details of programme design, decoupled payments may increase overall net returns to agriculture and/or shift the distribution of net returns in favour of smaller, more marginal farms. This would tend to keep more land (including more marginal land) in production. Decoupled payments may thus affect individual producers' decisions to exit farming and influence whether their land and other resources are withdrawn from production or simply transferred to other producers and/or other commodities. Evidence suggests that the number of farmers in the OECD countries is falling, but that the level of resources devoted to agricultural production is not.

Given the limitations of the AMS in measuring actual levels of support to agriculture and the conceptual and empirical difficulties associated with assessing the impact of decoupled payments on production and trade, considerable uncertainty surrounds the potential impact of further domestic support disciplines currently being negotiated in the Doha Round. FAO has highlighted elsewhere (FAO, 2005b) the need for a number of issues to be addressed:

- Criteria for the categorization of policies as exempt from reduction, particularly

those classified as decoupled, require effective review and clarification.
- Mechanisms to allow the reallocation of support across the different categories or boxes need to be established in a way that facilitates the shift towards less trade-distorting support but prohibits the exemption of policies that are, in effect, trade-distorting.
- Weaknesses in the way domestic support is currently measured in the WTO should be reviewed to ensure that further disciplines are effective.

Export competition[4]

The second of the three pillars of the AoA dealt with export competition. Although the original GATT 1947 prohibited the use of export subsidies in most sectors, an exception was made for primary products, including agricultural products. Export subsidies were prohibited in the manufacturing sector because they permit goods to be sold at less than the cost of production in the home country, a practice known as "dumping", which was agreed to constitute unfair competition. The AoA sought to redress this omission by establishing disciplines on export subsidies and other forms of export competition.

Under the Agreement, export subsidies had to be notified to the WTO and new measures of this type were prohibited. In addition, the budget outlay on export subsidies and the volume of subsidized exports were capped and reductions were required during the implementation period. The AoA also required Members to negotiate disciplines on the use of export credit guarantees and food aid shipments that might be used to circumvent the disciplines on direct subsidies.

The WTO Framework Agreement calls for the development of modalities that will ensure the parallel elimination of all forms of export subsidies and disciplines on all export measures with equivalent effect. While there is little disagreement on proceeding with negotiations along these lines, determining "equivalent effects" is not a simple task. There is a danger that some policy instruments that have little effect on world market conditions in comparison with their potential benefits will be disciplined too stringently.

Three broad components of export competition are the focus of the current negotiations: (i) policies in direct support of an exported commodity, such as export subsidies and officially supported export credits; (ii) interventions in support of state trading enterprises; and (iii) food aid, notably that component of food aid used to facilitate the disposal of a country's surplus production.

Incidence of direct export subsidies

Of the 21 WTO Members that have the right to use export subsidies under the AoA, nine currently use them.[5] Of these countries or groupings, the EU is dominant, accounting for 90 percent of the value of export subsidies notified to the WTO during the period 1995–2001. Switzerland follows, with 5.3 percent of the total, and Norway and the United States each account for 1.4 percent. The use of export subsidies has declined significantly over the past decade – from some $7.5 billion in 1995 to less than $3 billion in 2001. The reductions observed in the EU have occurred not just as a reflection of meeting commitments under the AoA (given that the EU has not reached close to its ceiling for most commodities), but as a result of parallel domestic policy reform that has reduced, for many products, the need for such extensive use of export subsidies. As Figure 13 clearly illustrates, however, some EU exports are far more dependent on export subsidies than others. It should also be noted that the proportion of EU sugar exports that benefit from export subsidies is disputed.

---

[4] The material in this section draws on FAO (2005c).

[5] The EU (including Cyprus, the Czech Republic, Hungary, Poland and Slovakia), Israel, Mexico, Norway, Romania, Switzerland, Turkey, the United States and the Bolivarian Republic of Venezuela. Notification data generally lag by a few years; for some of the listed countries the most recent data are for 1998.

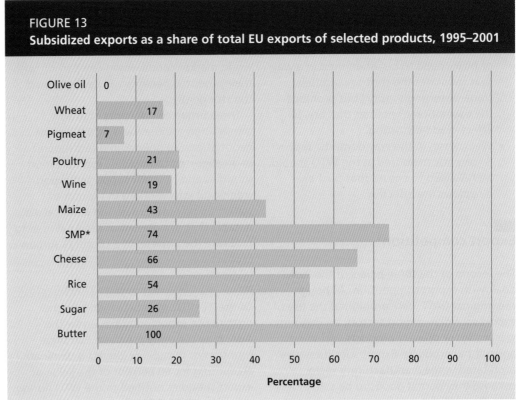

FIGURE 13
Subsidized exports as a share of total EU exports of selected products, 1995–2001

Product	Percentage
Olive oil	0
Wheat	17
Pigmeat	7
Poultry	21
Wine	19
Maize	43
SMP*	74
Cheese	66
Rice	54
Sugar	26
Butter	100

* Skimmed milk products.

Source: Jales, 2004.

Equivalence and incidence of indirect export subsidies

The equivalence of indirect export subsidies with direct export subsidies is usually discussed in terms of the effect of a given policy or activity on transactions and trade flows, or in terms of the gross expenditure on that policy or activity. Alternative approaches to analysis of the market effect of indirect subsidies include the extent of cost reduction (i.e. the reduction in cost to a foreign buyer relative to the domestic buyer of the commodity) and, related to this, the budgetary transfer involved in disposing of the commodity.

Export credits

The OECD (2000a) has attempted to determine the subsidy equivalence of export credits by country. This study defined export credit as "a guarantee, insurance, financing, refinancing or interest rate support arrangement provided by a government which allows a foreign buyer of exported goods and/or services to defer payments over a period of time".

Information on the incidence of the use of export credits is extremely difficult to obtain given that countries are not currently obliged to notify their use of such expenditure to the WTO and the terms under which export credits are provided are deemed to be of a confidential nature. Most analyses and viewpoints are based upon information presented by the OECD and using data from the period 1995–98 only.

In aggregate, export credits increased from $5.5 billion in 1995 to $7.9 billion in 1998. The majority of export credits and fully 95 percent of long-term credits were used by the United States. In the EU, the other significant user, the level of export credits was significantly lower than the use of export subsidies.

The OECD estimates of the subsidy equivalent of export credits provided by different countries take into account a number of factors related to repayment terms (interest rate, repayment period, etc.). For three OECD countries (Australia, Canada and the United States), the subsidy elements of export credit operations were higher than their export subsidy expenditures. The OECD found that the export credits from the United States in 1998 had a higher per unit subsidy equivalent than those from other countries, mainly by virtue of their longer repayment terms. Even so, the subsidy equivalent indicates that the importers paid, on average, 6.6 percent less for those transactions that were facilitated by United States export credits than they would have done without access to this support. These numbers are corroborated by estimates of about 9.9 percent from the United States General Accounting Office.

Given the relatively small export subsidy component of these export credits, which essentially focus on the "price" element of the credit (i.e. how much cheaper they make the exports compared with commercial alternatives not benefiting from credits), perhaps a more relevant issue relates to how sensitive export patterns are to the use of credits. A key question for further research is whether the removal of credits associated with long-term trade arrangements will cause a switch in the sourcing of the commodity away from the country previously extending the credit. This would depend on the elasticity of substitution of an importing country's imports from different countries, with higher elasticities implying greater scope for substitution.

State trading enterprises

As in the case of export credits (and unlike the case of export subsidies), there is an unresolved debate as to the balance of the relative merits and costs of the existence of state trading enterprises (STEs). On the one hand, such entities have been criticized in relation to their use of their monopoly status to influence market conditions and trade flows, and for the lack of transparency regarding their actions. For example, they may be granted subsidies by governments to facilitate their operations at below cost. Against this, others argue that STEs are a useful response to imperfect world commodity markets. Activities such as price pooling and the underwriting of losses, which can produce similar effects to those of export subsidies, can also be beneficial in reducing risks to farmers and traders (Young, 2004a). In addition, their large size (in terms of the volumes transacted) allows them to compete with large multinational trading companies, whose own use of market power has attracted criticism.

In determining whether, and indeed how, to eliminate or discipline certain actions undertaken by STEs, it is important to bear in mind these relative merits and to try to understand more clearly whether, on balance, the activities of individual STEs are detrimental (and should therefore be restricted) or beneficial (where more care might be required before attempting to restrict certain activities).

The OECD (2000b, 2000c) provides a comprehensive review of the existence and activities of STEs in OECD member countries. In developing countries, examples include China's COFCO, trading in cereals, oils and foodstuffs, and Indonesia's Bulog. However, the latter are believed to have limited market power. From a political point of view the perceived importance of a relatively small number of key STEs drives the argument for more stringent disciplines. These include the Australian Wheat Board and the Canadian Wheat Board, which together account for 40 percent of the global wheat market; the United States Commodity Credit Corporation; and Fonterra in New Zealand,[6] which accounts for 30 percent of global dairy exports (Young, 2004a).

From an empirical viewpoint, there is little evidence that the STEs cause significant market distortion. Sumner and Boltuck (2001) and Carter and Smith (2001) found no evidence of market power for the Canadian Wheat Board and no evidence that its actions harmed United States exporters. Indeed, there are no widely accepted studies indicating that existing STEs are currently distorting markets in a significant way. Concern remains, however, that these STEs could increase their activities, which would also raise their potential to create market distortion if their activities are not subject to discipline at the same time as other components of export competition.

Theoretical analysis can be used to gain insights into the potential distorting impact under a range of situations and to identify STE characteristics that may be more market-distorting than others. McCorriston and MacLaren (2004) attempted to operationalize a definition of subsidy

---

[6] The former New Zealand Dairy Board STE is now a farmer-owned cooperative, renamed Fonterra.

equivalence as "the export subsidy that would be paid to ... private firms to replicate the same quantity of exports that arise in a given STE environment". They found that a subsidy equivalent defined in this way can be positive or negative. Factors determining the impact of an STE include (i) how competitive the market would be in the absence of the STE and (ii) the actual configuration and actions of the STE – different STEs will not necessarily have the same magnitude of impact or even the same direction of impact in terms of trade distortion.

In terms of equivalence, evidence (both empirical and theoretical) suggests that an increase in export levels will always be higher with the use of direct export subsidies than if the same amount of support is provided via financial assistance to an STE. In relation to the insights arising from the theoretical framework proposed by McCorriston and MacLaren (2005), a number of observations can be made:

- *Competitiveness of the market.* There are widely held concerns about private exporters, given that the international trade of many agricultural commodities is concentrated in the hands of a few private multinational firms with the capacity to exert considerable market power. It is argued that international markets are far from being perfectly competitive and that private exporters compete with STEs in an oligopolistic market. Scoppola (2004) argues that there is, however, some debate as to whether either multinational firms or STEs can exert market power on, for example, international grain markets. Analyses by Caves and Pugel (1982), Carter, Loyns and Berwald (1998) and Carter and Smith (2001) suggest that they cannot. Others have argued that both can exert market power and are able to influence international prices in oligopolistic markets (e.g. Larue, Fulton and Veeman, 1999; McCorriston and MacLaren, 2002; Hamilton and Stiegert, 2002).
- *Exclusive rights vs ownership.* Theory suggests that the issue concerning competitive behaviour of STEs is not whether they are publicly or privately owned, but the nature of rights that they have to procure and to disburse products. Exclusive rights for exporting STEs can apply in both the domestic or export markets and/or apply both to sales and procurement. These rights differ across STEs. For example, the Canadian Wheat Board has exclusive rights in the domestic and export markets, while others only have exclusive rights in the domestic market. STEs and private firms can also differ with respect to their objective function. STEs often have a wider social mandate, for example in reducing consumer food prices or stabilizing producer prices, than that of private firms, which are concerned more with maximizing returns to stakeholders. A number of authors argue that this can result in significantly different trade impacts (e.g. Dixit and Josling, 1997; McCorriston and MacLaren, 2002; Carter, Loyns and Berwald, 1998; and Carter and Smith, 2001).

Food aid

Disciplines on mechanisms by which food aid is procured and/or disbursed are under negotiation primarily in response to fears that the use of food aid as a mechanism for surplus disposal will increase if countries become more constrained in their access to other mechanisms for supporting exports. However, food aid, by definition, is also a humanitarian issue and there are grave concerns that disciplining food aid in an indiscriminate manner, while reducing the scope for the use of forms of food aid that are potentially more distorting, will also have a negative impact on its beneficial aspects.

Food aid is disbursed in a number of forms that may displace commercial imports to different degrees. Food aid can be categorized as "emergency" or "non-emergency", with a number of subdivisions within the latter category. The view that emergency food aid should not be restricted is broadly supported because any commercial trade displacement or international market distortion resulting from emergency food aid is likely to be minimal. Emergency food aid accounts for around 60–70 percent of total food aid disbursement.

In the case of non-emergency food aid, there is some dispute about the impacts of different mechanisms for both procurement and disbursement. Non-emergency food aid can be divided into targeted food aid, which is given as food to recipients (examples include food-for-work or school lunch programmes) and monetized food aid, which is sold on local markets and the cash from its sale used to fund development projects.

The impact of food aid on markets is measured using the concept of additionality. Food aid is defined as additional if it is given to people who, because of their inability to access food by other means, would not have consumed the equivalent amount of food otherwise. Intuitively, emergency food aid should be closest to being fully additional in consumption as the recipients are, by definition, in distress and would not otherwise have access to alternative sources of food. Food aid that is wholly additional would have no distorting effects on production or commercial trade.

Although there are few empirical estimates of the additionality of monetized food aid, the extent of additionality is likely to be less than for targeted food aid and will depend upon how it is delivered. Against this, the benefits to recipients of, for example, agricultural development projects funded via the monetization of food aid need to be considered (Young, 2004b).

Additionality is likely to be situation-dependent. In conflict situations, the ability to import may otherwise be restricted and food aid would be expected to be more additional. Rates of inflation can also be high and wage earners unable to work in such situations – both factors contributing to the reduced ability of individuals to access alternative food sources (Young, 2004b). Additionality can also depend on programme design and implementation. The use of funds generated and whether they enhance demand or supply (i.e. whether they are used to increase direct consumption or to fund supply-enhancing agricultural projects) will contribute to the extent of additionality.

The way forward on export competition

The export competition issue is central to the ongoing round of trade talks. It is expected that direct export subsidies will be phased out eventually, along with the subsidy element in other export programmes. Moreover, certain practices have been challenged through the WTO dispute settlement process, putting further pressure on both the EU and the United States to make substantial reforms in this area.

Agreements in the WTO have generally been developed on the basis of simple rules, and not on the results of complex models. The measurement of equivalence, while conceptually feasible, is in practice likely to require sophisticated analysis to determine the relative effect of various components of export competition. To move the negotiations on export competition forward, it will be necessary to develop simple rules to discipline trade-distorting activities without removing the benefits that they provide in reducing market imperfections in, for example, capital markets, and their associated development and humanitarian benefits.

One general approach to developing such rules would be to group activities in terms of their likelihood to influence trade flows, not on the basis of their price equivalence, even where this could, in theory, be measured, because the latter would require a more complicated set of rules and criteria.

The combination of measures may matter more than their individual effects. Developing a workable grouping would therefore depend on how substitutable the practices are. If, at the extreme, they were perfectly substitutable it would be necessary to discipline them all. Evidence suggests, however, that this is not necessarily the case, and although some level of reinstrumentation could occur, stringent disciplines are likely to be inappropriate.

In considering the development of new rules on export competition, the form of WTO notifications will also be important. Decisions will need to be taken on which practices should be included in the notification obligations. Once decided, it will also be necessary to identify the information required in order to understand how these policies might work. To ensure workable disciplines and compliance, notifications would also need to be more timely than at present.

> **BOX 3**
> **The European Union's tariff rate quota regime for dairy products**
>
> The figures below provide an overview of the 1995 and 2000 allocation of dairy quotas by the EU. In both periods covered, roughly 95 percent of dairy imports, by value, were covered by TRQs. Several features are apparent. First is the complexity of the regime, which involves separate TRQs for skimmed milk powder, butter and five categories of cheese, with different quota levels, in-quota tariffs and out-of-quota tariffs for each category.
>
> The second feature is that in 1995 the in-quota tariffs for some product categories were so high that the import quota levels (based on current access commitments, or Uruguay Round base imports) were not even met, leading to an apparent erosion of market access since the Uruguay Round base period.
>
>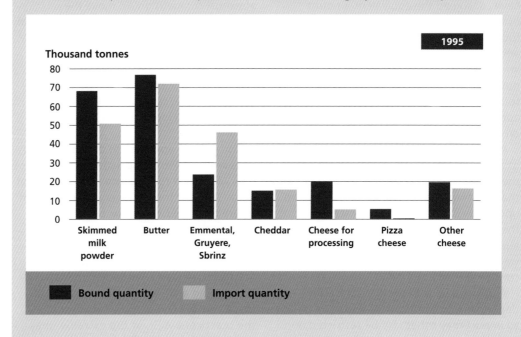

Market access

The market access provisions of the AoA are extremely complex because of the wide variety of market barriers imposed on agriculture prior to the negotiations and because of the critical role of market access in disciplining other forms of support to agriculture.

Many domestic agricultural policies and export subsidies cannot function without restrictions on market access. If a country is open to imports, there is a natural limit to the degree of support it can provide to its own farmers because farmers in other countries will quickly expand their exports to capture part of the support. The United States experienced this in the early years of the US-Canada Free Trade Agreement when its efforts to raise domestic barley prices through the use of export subsidies were met with a surge of barley imports from Canada (Haley, 1995).

A wide range of tariffs and non-tariff barriers (NTBs) such as import quotas and variable levies were applied to agricultural products before the Uruguay Round. The elimination of NTBs was a particular focus of the Uruguay Round negotiations because such barriers tend to distort markets more severely than an equivalent tariff and are less transparent. Unlike tariffs, NTBs block the transmission of price signals between the world market and domestic markets. This prevents domestic supply and demand

In contrast with 1995, in 2000 all quotas except for pizza cheese were exceeded, meaning that the binding constraint on further imports was the out-of-quota tariff (and that quota rents accrue on the in-quota amounts – roughly half of imports).

Because of the bilateral allocation of quotas, the system discriminates against third-country suppliers. For example, the full butter quota for 1995 was allocated exclusively to New Zealand, while the Cheddar cheese quota was shared by Australia, Canada and New Zealand. In 2000, all quotas were overfilled, though again with a bias towards those countries given import quotas and hence preferential access.

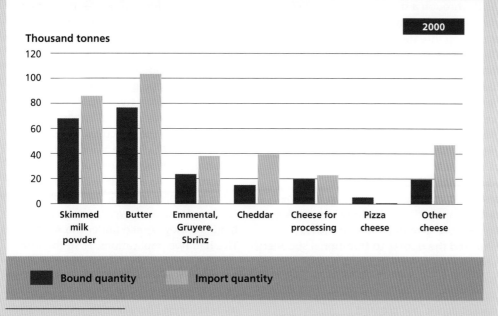

Source: Francois 2001a; AMAD database.

from adjusting in response to world market conditions and it shifts the burden for domestic market stabilization onto world markets (i.e. to countries that do not use such measures). Tariffs, in contrast, allow price signals to be transmitted more readily between world and domestic markets, reducing the distortion of world market prices.

During the negotiations, a variety of mechanisms were used to convert NTBs to tariffs and to reduce the resulting tariffs. The stated objective was to reduce the level of protection and the degree of trade distortion created by that protection. Some of the specific mechanisms employed and the rules on tariff reduction limited the amount of real market access liberalization that took place.

It can even be argued that some of the new mechanisms themselves constitute NTBs. The major criticisms of the Uruguay Round market access provisions focus on the mechanisms for converting NTBs to tariffs, the tariff-reduction formula and a perceived imbalance between the rights and obligations of developed and developing countries.

Under the Uruguay Round negotiations, developed countries agreed to convert their NTBs to equivalent tariffs through a process known as "tariffication", while developing countries were given the option of simply replacing their NTBs and unbound tariffs with bound tariffs, known as "ceiling bindings". The resulting tariffs were reduced on the basis of a simple unweighted average.

The tariffication process was meant to ensure that developed countries established tariffs that were no more trade-restrictive than the NTBs they replaced. Where tariffication was used, countries were required to introduce tariff rate quotas (TRQs) to ensure that effective market access was not eroded. TRQs involved commitments on current access, made in quantity terms, while some liberalization was to be guaranteed through the creation of minimum access commitments, set at 5 percent of 1986–88 consumption levels.

In addition, for tariffied commodities, countries could claim the right to increase tariffs through a special safeguard (SSG) mechanism if an import surge or sudden price drop threatened their producers. Thirty-eight WTO Members established TRQ commitments for a total of 1 379 quotas and claimed SSG privileges on 6 072 individual tariff items. Very few developing countries are among this group.

In practice, TRQs have done little to improve market access. The combination of current access and market access commitments has led directly to quantitative commitments (and in some cases quantitative restrictions) on market access. Furthermore, many countries allocated the quotas to traditional suppliers and counted pre-existing preferential access quotas as part of their minimum access commitments with the result that no new market access was created.

Unlike simple tariffs, TRQs generate market rents that may be captured by various groups (producers, exporting governments, importing governments and traders) depending on the administrative mechanism and the degree of market competition. It has been estimated that new access volumes created by TRQs typically accounted for less than 2 percent of world trade for the commodities in question, and TRQ utilization rates or fill rates have averaged only about two-thirds. Thus, TRQs have not been as effective in ensuring an increase in market access as expected. One example of the operation of TRQs is the EU's dairy policy, described in Box 3.

Most developing countries and LDCs chose the option of adopting tariff ceilings to replace their import quotas instead of going through the tariffication process (often declaring a single bound tariff rate for all agricultural commodities). Developing countries were also allowed to reduce their bound tariffs by smaller amounts than were the developed countries (24 percent versus 36 percent) and the LDCs were exempt from reduction commitments. These provisions were meant to provide special and differential treatment, but in practice they resulted in an imbalance between developed and developing countries that is arguably in favour of the former. Because most developing countries and LDCs did not tariffy they did not create TRQs and could not claim SSG privileges. Thus, bound tariffs are their only form of border protection. Because TRQs and SSGs are more trade-restrictive than tariffs, developed countries have retained more latitude to protect sensitive commodities.

Many developing countries and LDCs had already eliminated import quotas and substantially reduced import tariffs in the context of the structural adjustment programmes that were being undertaken simultaneously with the Uruguay Round negotiations. As a result, when the AoA came into force in 1995, their applied import tariffs were much lower than the tariff bindings they agreed under the Agreement. This had two implications. First, the AoA required relatively little reduction in applied tariffs for these countries. Second, they had already undertaken significantly greater market access liberalization under structural adjustment than was required under the AoA. Box 4 discusses the importance of tariff revenues for the fiscal budgets of many developing countries.

In addition to the problems created by the tariffication process, the Uruguay Round formula for tariff reduction limited the amount of actual market access that was achieved and further distorted markets. Because the tariff reduction commitments were based on a simple average, countries could strategically reduce tariffs on "sensitive" high-tariff products by the minimum amount (15 percent and 10 percent, respectively, for developed and developing countries) while reducing tariffs on less-sensitive products by greater amounts to reach the average requirement. As a result, many of the highest pre-Uruguay Round tariffs were reduced by the smallest

> **BOX 4**
> **Tariffs as tax revenue**
>
> One reason why tariff reductions concern many developing countries is their potentially negative impact on tax revenue. In more than 25 developing countries tariff revenue can exceed 30 percent of the government's total tax revenue. In high-income countries, tariff revenues typically represent less than 2 percent of total tax revenue.
>
> The WTO highlights two revenue implications of trade liberalization. First, trade liberalization that substitutes tariffs for non-tariff barriers (e.g. quotas and restrictive licensing requirements) may have a positive revenue impact. Second, once trade protection is based on tariffs, the revenue implications of reductions in applied rates depend on the price elasticity of imports. Simulations suggest that price elasticities in open economies have to be much higher than empirically observed elasticities for trade liberalization to be self-financing (Devarajan, Go and Li, 1999). These findings imply that significant tariff reductions should be accompanied by reform of the general tax system to avoid the emergence of fiscal deficits or curtailment of government expenditure (Ebrill, Stotsky and Gropp, 1999).
>
> On the other hand, the empirical evidence on the impact of major trade liberalization programmes to date shows that revenue implications are not necessarily significant. For Bangladesh, Chile and Mexico, trade liberalization since the mid-1980s involved cuts in applied tariffs of more than 10 percentage points, reducing the ratio of duties to total tax revenue significantly in Bangladesh, but only slightly in Chile and Mexico. In each case, import growth accelerated sharply. Interestingly, in the initial years of trade liberalization in Chile and Mexico, the ratio of import duties to total tax revenue rose, but declined steadily thereafter.
>
> *Source:* WTO, 2003.

amounts, while already-low tariffs were reduced more. This created little new market access and increased the dispersion of the tariff rates of many countries, arguably increasing the distorting effect of tariffs on their markets.

Tariff escalation is a particular type of tariff dispersion that is of special importance to developing countries. It occurs when tariff levels increase with the degree of processing of a product. This favours imports of raw materials and discourages local processing in the exporting country. As developing countries attempt to add value to their agricultural products and take advantage of greater returns to differentiated value-added goods, tariff escalation works against their efforts. Given the higher income elasticity of demand for processed products, the impact of tariff escalation on the production and trade of processed products and on rural employment could be significant.

Tariff escalation is particularly pronounced in agriculture, with processed agricultural products being subject to significantly higher tariffs than raw farm products. Figure 14 shows most-favoured nation (MFN) tariffs for plant- and animal-based fibres (basic raw materials), textiles (intermediate goods) and clothing (a final good at the end of the processing chain). For these products, tariff escalation exists in both rich and poor countries. The relative gap is often higher in the OECD countries, though the absolute gap can be very high for developing countries also.

The fact that the developed countries' tariff structures protect the market for processed products more than they do for primary products is seen as an obstacle for the industrial and economic development of developing countries (FAO, 2004a). Many developing economies also tend to apply systematic tariff escalation and high tariffs to the final stage of processing. Bangladesh and Morocco, for example, both engage in far greater absolute tariff escalation than do the OECD countries. High absolute levels

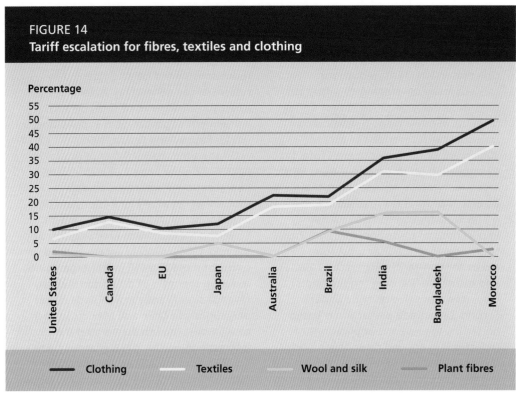

FIGURE 14
Tariff escalation for fibres, textiles and clothing

Source: Comtrade and GTAP v. 6.4 databases.

of tariff escalation in developing countries suggest that potentially large gains could be realized if escalation were removed by developing economies themselves (Rae and Josling, 2003).

Measuring agricultural protection

Given the complexity of the market access commitments made in the Uruguay Round, their importance in facilitating the use of domestic and export subsidies, and their prominence in the Doha Round of negotiations, this section explores the measurement of market access barriers in greater detail.

Measuring the extent of agricultural protection may seem simple, as tariff schedules typically provide information at a high level of detail. However, there are difficulties involved, not least because of the differences between bound rates (the policy variable considered in WTO schedules of concessions) and applied tariff rates. Complications also arise when aggregating from the fine level of detail in tariff schedules up to the broader commodity aggregates that allow an overall evaluation and comparison with protection regimes in other countries. This analysis attempts to take many of these complexities into account.

Table 5 summarizes a market access data set consisting of 65 305 tariff lines at the six-digit level of the Harmonized System for 103 countries for the period 2000–02.[7] It incorporates *ad valorem* equivalents for tariffs that include a specific element. This is important because, as shown in Messerlin (2003) and World Bank (2005a), these *ad valorem*-equivalent specific tariffs are frequently much higher than the *ad valorem* tariffs alone. As the primary focus of current WTO negotiations is on increasing market access rather than the redistribution of quota rents, in-quota tariffs for tariff rate quotas are excluded from the analysis.

---

[7] This data set was compiled by Martin and Zhi (2005), from two major sources: the UNCTAD/TRAINS database and a dataset developed at the United States Department of Agriculture (USDA) (Wainio, Gibson and Whitley, 2001; Wainio and Gibson, 2004).

TABLE 5
Country-level agricultural tariff data, 2000–02

Countries	Simple average		Coefficient of variation		Weighted average		Binding overhang	Tariff lines		Maximum rate	
	Applied	Bound	Applied	Bound	Applied	Bound		Bound at zero	Total	Applied	Bound
	(Percentage)		(Percentage)		(Percentage)		(Percentage of bound rate)	(Number)		(Percentage)	
INDUSTRIAL COUNTRIES											
Australia	1.3	3.2	176.9	143.8	2.4	4.9	51.0	224	724	13	29
Canada	9.8	14.1	266.3	308.5	11.7	17.1	31.6	267	636	161	620
European Union	19.8	22.5	157.6	167.6	17.4	21.3	18.3	152	604	327	479
Iceland	47.8	114.7	218.4	139.1	24.5	60.9	59.8	115	617	584	963
Japan	24.2	48.4	269.8	281.6	20.9	51.6	59.5	179	613	716	1 646
New Zealand	1.6	5.9	143.8	122.0	2.4	8.0	70.0	342	685	7	31
Norway	83.2	168.6	219.2	126.4	36.4	116.4	68.7	126	648	3 424	3 424
Switzerland	28.1	51.7	198.2	138.9	21.2	44.2	52.0	77	572	646	666
United States	5.0	6.1	220.0	203.3	5.0	6.6	24.2	170	596	97	100
All industrial countries	**24.1**	**47.7**	**336.3**	**246.3**	**14.1**	**24.9**	**43.4**	**1 652**	**5 695**	**3 424**	**3 424**
DEVELOPING COUNTRIES											
East Asia and the Pacific	17.0	48.6	380.0	286.4	39.1	59.4	34.2	112	4 466	2 565	7 696
China	15.7	15.8	72.0	72.8	12.6	12.8	1.6	18	670	65	65
Indonesia	7.5	46.8	261.3	46.4	3.2	54.8	94.2	–	734	150	210
Republic of Korea	54.7	64.9	228.2	197.4	103.7	112.9	8.1	11	563	800	887
Malaysia	11.8	35.6	998.3	950.8	29.2	86.6	66.3	79	594	2 565	7 696
Myanmar	8.6	103.1	91.9	90.8	10.5	141.3	92.6	4	631	40	550
Papua New Guinea	17.6	43.3	103.4	49.2	8.1	34.6	76.6	–	607	75	100
Philippines	9.3	34.7	114.0	32.9	8.3	29.9	72.2	–	667	58	80
Europe and Central Asia	13.9	29.1	127.1	140.7	15.8	51.1	69.1	412	6 429	336	336
Albania	9.4	9.4	58.5	58.5	9.5	9.5	0.0	73	671	20	20
Armenia	7.1	14.8	64.8	8.8	6.6	15.0	56.0	3	671	10	15
Bulgaria	18.0	35.8	81.7	75.7	20.6	33.5	38.5	34	577	74	98
Croatia	8.3	9.4	80.7	95.7	9.3	10.5	11.4	104	605	25	44
Estonia	11.5	17.6	133.9	80.7	7.6	13.4	43.3	115	671	59	59
Kyrgyzstan	8.4	12.4	51.2	38.7	8.6	11.7	26.5	8	657	18	25
Latvia	11.3	34.8	115.0	53.7	9.9	23.7	58.2	14	667	50	55
Lithuania	9.1	15.4	153.8	92.2	9.0	13.1	31.3	55	666	87	100
Romania	24.1	99.1	94.2	83.5	32.0	141.5	77.4	1	671	248	333
Thailand	34.8	43.0	94.8	81.4	15.3	51.4	70.2	5	573	336	336
Latin America and Caribbean	13.4	59.2	92.2	64.1	18.4	51.8	64.5	55	18 726	254	257
Argentina	12.1	32.3	41.3	23.2	13.7	31.1	55.9	2	734	22	35
Belize	16.7	101	99.4	3.9	12.9	100.8	87.2	–	606	110	110
Bolivia	10.0	40.0	8.0	1.0	9.9	40.0	75.3	–	734	17	40
Brazil	12.2	35.5	42.6	28.2	11.5	42.5	72.9	14	734	44	55
Chile	7.9	25.7	3.8	7.8	8.0	26.3	69.6	–	734	9	32
Colombia	14.8	91.6	35.1	36.6	14.6	112.3	87.0	–	734	20	227
Costa Rica	11.8	42.1	120.3	56.1	10.8	33.5	67.8	–	734	99	233
Cuba	9.8	36.9	77.6	28.5	10.0	31.1	67.8	31	671	30	40
Dominica	19.2	112.8	125.5	19.2	22.3	125.4	82.2	–	649	140	150
Dominican Republic	15.7	40.0	61.1	0.0	12.5	40.0	68.8	–	641	38	40
Ecuador	14.6	25.3	36.3	36.8	14.3	26.7	46.4	–	551	20	72
El Salvador	10.8	42.0	83.3	46.7	12.5	43.6	71.3	–	734	40	164
Grenada	16.0	101.2	90.0	33.2	15.0	82.7	81.9	5	602	40	200

TABLE 5 (cont.)

Countries	Simple average		Coefficient of variation		Weighted average		Binding overhang	Tariff lines		Maximum rate	
	Applied	Bound	Applied	Bound	Applied	Bound		Bound at zero	Total	Applied	Bound
	(Percentage)		(Percentage)		(Percentage)		(Percentage of bound rate)	(Number)		(Percentage)	
Guatemala	9.9	49.8	74.7	79.9	10.9	63.8	82.9	–	733	33	257
Guyana	17.6	100.0	96.6	0.0	18.0	100.0	82.0	–	605	100	100
Honduras	10.2	32.2	72.5	21.7	10.6	28.2	62.4	–	734	55	60
Jamaica	15.5	100.0	109.0	0.0	16.4	100.0	83.6	–	648	75	100
Mexico	20.9	41.1	123.4	71.8	28.2	51.8	45.6	1	599	254	254
Nicaragua	8.1	40.4	87.7	6.9	11.1	41.9	73.5	–	606	53	60
Panama	12.8	27.4	103.1	51.8	11.7	22.2	47.3	2	626	144	144
Paraguay	11.6	35.0	39.7	0.0	16.2	35.0	53.7	–	649	31	35
Peru	17.2	30.9	38.4	17.8	16.5	40.1	58.9	–	577	30	68
Saint Kitts and Nevis	14.0	108.8	111.4	26.7	18.1	98.1	81.5	–	602	130	250
Saint Lucia	14.2	114.4	104.2	23.1	15.5	116.7	86.7	–	605	45	250
Saint Vincent	15.4	114.8	93.5	23.0	15.9	115.0	86.2	–	602	40	250
Suriname	11.4	19.9	65.8	3.5	13.2	19.9	33.7	–	343	20	20
Trinidad and Tobago	14.5	100.2	109.7	3.3	13.9	100.0	86.1	–	604	70	156
Uruguay	12.3	33.9	39.8	21.2	13.9	33.1	58.0	–	671	30	55
Venezuela (Bolivarian Republic of)	14.8	55.5	35.1	60.9	16.2	74.2	78.2	–	664	20	135
Near East and North Africa	**31.0**	**61.0**	**124.1**	**297.4**	**22.4**	**50.0**	**55.2**	**6**	**4 039**	**600**	**3 000**
Djibouti	20.5	47.5	56.6	85.9	18.5	54.1	65.8	–	647	40	450
Egypt	21.8	96.0	122.5	448.3	6.3	23.6	73.3	–	661	600	3 000
Jordan	20.1	23.9	123.9	129.3	13.8	18.4	25.0	6	667	180	200
Morocco	41.0	54.6	100.2	91.6	27.0	81.9	67.0	–	734	289	289
Oman	11.0	28.3	208.2	161.5	39.9	66.1	39.6	–	663	100	200
Tunisia	70.0	115.9	75.6	35.0	46.9	75.2	37.6	–	667	200	200
South Asia	**23.0**	**100.9**	**60.1**	**66.5**	**22.3**	**132.4**	**83.2**	**10**	**3 129**	**150**	**300**
Bangladesh	23.5	187.8	57.4	22.7	14.3	160.2	91.1		635	38	200
India	35.3	114.8	52.7	47.3	28.4	147.2	80.7	10	621	150	300
Maldives	18.4	48.5	40.8	139.4	16.9	66.4	74.5	–	624	50	300
Pakistan	18.4	100.1	44.6	10.1	12.6	109.0	88.4	–	648	30	150
Sri Lanka	19.2	50.0	53.1	0.0	16.2	50.0	67.6	–	601	50	50
Sub-Saharan Africa	**17.5**	**74.6**	**75.0**	**53.6**	**16.2**	**73.5**	**78.0**	**78**	**17 117**	**133**	**200**
Angola	9.4	52.8	87.2	17.8	13.0	49.3	73.6	–	668	35	55
Benin	13.9	61.4	48.2	19.7	14.9	54.8	72.8	–	671	20	100
Burkina Faso	13.9	98.1	48.2	12.7	14.0	81.4	82.8	–	671	20	100
Burundi	31.6	95.4	42.7	20.4	29.3	84.4	65.3	15	623	40	100
Cameroon	22.1	80.0	43.4	0.0	18.4	80.0	77.0	–	631	30	80
Central African Republic	22.1	30.0	43.0	0.0	23.7	30.0	21.0	–	667	30	30
Chad	22.1	80.0	43.4	0.0	25.6	80.0	68.0	–	631	30	80
Congo	22.1	30.0	43.4	0.0	23.5	30	21.7	–	631	30	30
Côte d'Ivoire	10.9	14.9	41.3	34.9	9.7	14.7	34.0	1	671	20	64
Gabon	22.1	60.0	43.0	0.0	22.2	60.0	63.0	–	667	30	60
Guinea-Bissau	13.8	40.0	48.6	0.0	17.4	40.0	56.5	–	626	20	40
Kenya	20.3	100.0	55.7	0.0	25.0	100.0	75.0	–	625	100	100
Madagascar	5.8	30.0	84.5	0.0	3.8	30.0	87.3	–	671	20	30
Malawi	15.1	121.5	60.9	13.3	14.1	118.6	88.1	–	635	25	125
Mali	13.9	59.2	48.2	11.8	13.5	54.2	75.1	–	671	20	75
Mauritania	12.6	37.7	60.3	44.6	8.0	43.9	81.8	–	671	20	75
Mauritius	18.6	119.3	124.2	11.8	12.7	96.9	86.9	–	578	80	122

TABLE 5 (cont.)

Countries	Simple average		Coefficient of variation		Weighted average		Binding overhang	Tariff lines		Maximum rate	
	Applied	Bound	Applied	Bound	Applied	Bound		Bound at zero	Total	Applied	Bound
	(Percentage)		(Percentage)		(Percentage)		(Percentage of bound rate)	(Number)		(Percentage)	
Mozambique	17.2	100.0	66.3	0.0	13.0	100.0	87.0	–	689	30	100
Niger	13.9	83.4	48.2	75.9	13.3	68.5	80.6	–	671	20	200
Nigeria	39.0	150.0	58.5	0.0	29.1	150.0	80.6	–	626	133	150
Rwanda	12.2	74.2	73.0	25.1	10.7	64.9	83.5	17	626	25	80
Senegal	14.0	29.8	47.9	5.0	11.5	28.3	59.4	–	671	20	30
South Africa	10.3	35.5	118.4	85.9	8.9	38.7	77.0	45	252	55	160
Togo	13.9	80.0	48.2	0.0	11.8	80.0	85.3	–	635	20	80
Uganda	12.6	77.7	28.6	10.2	9.3	78.5	88.2	–	698	15	80
Zambia	20.6	123.2	75.2	9.5	17.5	117.0	85.0	–	622	125	125
Zimbabwe	28.9	145.6	70.6	15.9	21.0	141.1	85.1	–	619	100	150
High-income non–OECD countries	**14.4**	**57.8**	**499.4**	**238.6**	**61.8**	**79.6**	**22.4**	**61**	**6 267**	**3 788**	**8 334**
Antigua and Barbuda	14.6	105.1	91.1	17.0	20.3	107.2	81.1	–	648	40	220
Bahrain	8.0	37.7	188.8	53.6	11.0	42.2	73.9	–	624	125	200
Barbados	25.6	111.0	127.7	22.3	33.0	108.8	69.7	–	654	163	223
Brunei	14.9	54.5	1 249.0	748.3	33.7	96.7	65.1	–	600	3 788	8 334
Cyprus	21.8	59.0	156.9	49.2	23.3	98.2	76.3	6	336	245	245
Kuwait	1.7	100.0	517.6	0.0	5.1	100.0	94.9	–	631	100	100
Malta	2.7	33.8	148.1	55.0	2.3	29.9	92.3	16	231	16	88
Qatar	4.9	26.3	159.2	163.9	6.6	26.5	75.1	–	629	70	200
Singapore	0.0	9.5	0.0	21.1	0.0	8.9	0.0	24	710	0	10
Slovenia	11.3	23.5	102.7	56.2	14.0	22.0	36.4	4	641	45	45
All developing countries	**16.3**	**61.7**	**189.9**	**136.7**	**24.4**	**60.0**	**59.3**	**723**	**59 610**	**3 788**	**8 334**
Upper middle-income countries	**13.7**	**56.5**	**211.5**	**146.1**	**23.1**	**54.1**	**57.3**	**377**	**13 541**	**2 565**	**7 696**
Lower middle-income countries	**18.0**	**51.4**	**122.4**	**176.6**	**14.4**	**41.8**	**65.6**	**230**	**19 043**	**600**	**3 000**
Low-income countries	**17.0**	**75.7**	**80.6**	**64.2**	**15.5**	**95.6**	**83.8**	**55**	**20 759**	**150**	**550**
WORLD	**17.0**	**60.5**	**224.2**	**145.1**	**18.0**	**38.2**	**52.9**	**2 375**	**65 305**	**3 788**	**8 334**

Source: Martin and Zhi, 2005.

Simple average tariffs

The first two columns of Table 5 present the simple average applied and bound agricultural tariffs by country and by regional and economic groupings. The country-group averages are calculated by weighting each country's simple average tariff by the size of its total agricultural imports, to allow for the fact that some economies are much larger than others. Several observations can be made from an examination of simple average tariffs.

First, it appears that simple average applied tariffs are higher in industrial countries (24 percent) than in developing countries (16 percent). This may be misleading because of the exclusion of in-quota tariffs on products subject to TRQs. TRQs are much more prevalent in the industrial countries, and the in-quota tariffs on these products are, on average, about half the rate of out-of-quota tariffs (Wainio, Gibson and Whitley, 2001).

Second, there is a striking degree of variation within both the industrial country group and the developing country group (countries classified as developing by the WTO). In some industrial countries, such as Australia and New Zealand, average applied tariffs are less than 2 percent. At the other extreme, Norway has an average of more than 80 percent. Within the developing country group, most countries have average applied rates of between 5 and 25 percent, although a few countries such as Tunisia (70 percent), the Republic of Korea (55 percent), Morocco (41 percent), Nigeria (39 percent), India (35 percent) and Thailand (35 percent) have substantially higher average rates.

Third, simple average bound rates appear to be much higher than applied rates, both in industrial and developing countries. For the industrial countries, the average bound rate of 48 percent is almost twice as high as the average applied rate. For the developing countries, the average bound tariff of 62 percent is more than three times the applied rate of 16 percent. Average bound rates are much higher for developing countries as a group, partly because these countries made more use of the option to bind tariffs using ceiling bindings in the Uruguay Round (Hathaway and Ingco, 1996). South Asia has the highest average bound tariffs, at more than 100 percent, with sub-Saharan Africa having the second-highest, at 75 percent.

Tariff dispersion
The trade-distorting effect of a tariff regime is influenced by both the average level of tariffs and the dispersion of tariff rates around the average. The coefficient of variation (CV) measures the dispersion or variability of tariffs relative to the mean. A tariff schedule that applies the same tariff rate to all products has a CV of zero. While a flat tariff schedule may discourage trade, depending on the level of the tariff, it does so equally for all products; therefore, it is less trade-distorting than is a tariff schedule having a high degree of dispersion.

The CVs of the industrial countries and the developing countries differ considerably. The variation of tariffs is typically much higher in industrial countries than in developing countries, with the CV for applied tariff rates in the industrial countries averaging 336 percent, as against 190 in developing countries. For bound rates, the difference is similarly striking, with the industrial country CV of 246 percent being almost twice the corresponding value of 137 in developing countries.

Among the developing countries, the higher-income countries have significantly higher tariff CVs than those of the low-income countries. In low-income countries, the CV of applied tariffs is generally less than 100 percent. Bound tariffs in developing countries are typically much less variable than applied rates, with some African countries having completely uniform tariff bindings indicated by CVs of zero.

Weighted average tariffs
Simple average tariffs give equal weight to all tariff lines and thus may be overly influenced by tariffs on unimportant items. Weighting tariffs according to the product's importance in trade can provide a more representative picture of a country's tariff schedule. Trade-weighting can introduce a downward bias, however, if some tariffs are so high that they eliminate trade altogether. With this caveat in mind, trade-weighted applied and bound tariffs are shown in the fifth and sixth columns of Table 5.

The weighted average tariff rates present a different picture than do the simple averages. The weighted average applied tariff is 14 percent in the industrial countries – well below the simple average of 24 percent. This is partly because many of the peak tariffs in industrial countries are so high that they restrict imports to very low levels, thus giving them too little weight in the average and underestimating their actual trade restrictiveness. For the developing countries, the opposite pattern emerges: the weighted average applied rate, at 24 percent, is above the simple average rate of 16 percent. Tariffs are less variable in developing countries and there are fewer mega-peak tariffs that effectively eliminate imports of the goods to which they are applied. The existence of these mega-peak tariffs in the industrial countries highlights the importance of ensuring that future tariff reductions bring about reductions in the highest tariffs.

Binding overhang
Another important factor to consider is the gap between bound and applied tariffs, or "binding overhang" (Francois, 2001b;

Francois and Martin, 2004; Francois, van Meijl and van Tongeren, 2005). Because negotiated tariff reductions generally involve bound tariffs rather than applied rates, a large overhang implies that even deep reductions in bound rates may lead to little actual liberalization. The measure of binding overhang is expressed using weighted average tariff data. The results in Table 5 are presented as percentages of the initial bound rate, providing an indication of the extent to which average bound rates would need to be cut to bring about substantial improvements in market access.

These data point to very high levels of binding overhang in both industrial and developing countries. In the industrial countries, the average binding overhang for agriculture is 43 percent. The 60 percent overhang in Japan inflates this figure. While discussions of binding overhang frequently emphasize developing countries, this result makes it clear that, at least in agriculture, the issue is also of importance in the industrial countries.

Nevertheless, the results confirm that the extent of binding overhang is greater in developing countries than in the industrial countries. The average in these countries is 59 percent. All income groups have binding overhang above 50 percent, except for the high-income group, where it is 22 percent. The East Asia region is the only developing country region where binding overhang is below 50 percent. In South Asia, however, it is an extraordinary 83 percent.

Yet another area where there are sharp differences between industrial and developing countries is in the share of tariff lines bound at zero. In the industrial countries, 29 percent of all tariff lines (at the six-digit level) are bound at zero, compared with 1.2 percent for developing countries. Among the developing countries, only those in Central Asia and Europe have any significant proportion of their tariffs bound at zero.

The last two columns in Table 5 show the maximum applied and bound tariff rates. The data indicate just how high the tariff peaks are in some countries, even when – as in this table – the tariffs analysed are at the six-digit level. While some of these peaks are on minor products, others are on potentially important products whose imports are tightly restricted.

Key findings

Governments have long intervened in food and agricultural markets, and although their policy objectives and tools have changed over time, they continue to view the sector as a vital policy domain. Until the Uruguay Round brought agriculture into the multilateral trading system, no internationally agreed rules existed to guide agricultural policy. The Uruguay Round AoA initiated a reform process in agriculture that is far from complete.

- Although many countries have redesigned their domestic agricultural support programmes to provide less-distorting forms of support, the overall level of support remains high, particularly in wealthier countries. The degree to which currently exempt forms of domestic support are decoupled from production continues to be debated, but the evidence suggests that some measures are less production-neutral than others.
- Export competition remains a contentious issue. While it may be possible to establish equivalence between export subsidies and other export competition measures at a conceptual level, caution should be exercised to avoid creating unnecessarily complicated disciplines. Further disciplines on food aid should weigh any potential market displacement effects against its humanitarian role.
- Tariff levels and other market access barriers remain high for agricultural products in both developed and developing countries. Prohibitively high tariff peaks and tariff escalation create severe distortions that systematically work against the efforts of producers in developing countries to enter the rapidly growing markets for processed products.
- Finally, the three pillars of the AoA are interlinked. Many developing countries will resist reducing their tariffs as long as their farmers must compete with subsidized production from other countries.

4. Macroeconomic impacts of agricultural trade reform

The agricultural and trade policy landscape described in Chapter 3 is complex. So too are the proposals for their reform. Assessing the economic effects of these proposed reforms is equally complex.[8] Simply observing the situation before and after a policy change is not sufficient to understand its impact. In reality, many changes – for example relating to other policies, the weather, technology – happen at the same time so the effects of any specific policy change can be difficult to disentangle.

Sophisticated econometric techniques are used to isolate the effects of policy changes *ex post*, or after the fact. But policy-makers often need to understand the potential impacts of alternative policy options *ex ante*, before they occur. *Ex ante* assessments of such options help identify potential winners and losers and aim to inform the policy debate. This chapter is concerned particularly with *ex ante* assessments of agricultural policy changes against the background of the ongoing Doha Round of multilateral trade policy negotiations.

The first section of this chapter describes some of the modelling approaches used in *ex ante* policy assessments, explaining their strengths and limitations. Several of the more recent attempts to predict the economy-wide impact of agricultural trade policy liberalization are discussed in the second section. More detailed commodity market impact studies based on agriculture sector models are described in the third section.

These modelling approaches yield some general indications about the likely winners and losers in the reform process at the national level, but they are less helpful in describing the distributional effects within a country. Chapters 5 and 6 extend the analysis to the household level, examining the impacts of trade policy reform on poverty and food security.

Modelling trade policy reform

Ex ante policy assessment is concerned with evaluating a situation with a proposed policy change against a situation without the policy change; economists therefore use models that simulate the structure of the economy and the ways in which different economic agents respond to policy changes. Economic models start from a portrait of an existing situation, and then proceed to paint a counterfactual world that includes the proposed policy changes.

The most commonly used models are market equilibrium models, containing equations that represent the responses of buyers and suppliers to changes in prices. Demand and supply are specified as functions of income, prices and elasticities. Prices adjust until markets clear, with demand and supply in equilibrium.

The behavioural response of suppliers and buyers is typically derived from optimizing assumptions. For a given production technology, the suppliers choose a combination of inputs such that costs are minimized for a given level of output. For a given set of consumer preferences, the buyers determine the combination of items that maximizes their utility for a given level of expenditure. These models typically assume constant returns technology, homothetic preferences (i.e. demand does not depend on the distribution of income), and markets characterized by perfect competition.

Depending on assumptions made about the flexibility of production factors (e.g. land, labour and capital), market equilibrium models can be classified as short-term, medium-term or long-term. In short-term

---

[8] This chapter is based on background papers prepared by van Tongeren (2005) and Francois, van Meijl and van Tongeren (2005).

models, some production factors are fixed, i.e. they are not allowed to move between alternative uses. Capital and agricultural land are usually held fixed in short-term models and agricultural labour is sometimes fixed. As the time frame of the model is extended, factors of production are gradually allowed to shift between uses. In long-term models, most factors can move between alternative uses.

Market equilibrium models can be further classified as partial or general equilibrium models, depending on whether they attempt to depict a single sector of the economy or the economy as a whole.

Partial equilibrium trade models treat individual international markets for selected traded goods. Such models for agricultural trade generally focus on trade in primary commodities. They capture agricultural supply, demand and trade for unprocessed or first-stage processed agricultural products without taking into account trade in processed food products, despite the fact that the latter commodities represent an increasing share of world trade. The economy-wide models, or general equilibrium models, attempt to account for the linkages with the rest of the economy.

The main area of application of partial equilibrium models is detailed trade policy analysis for specific products that represent a small portion of the economy in question. Policy-induced changes to a small sector are assumed to have little impact on the rest of the economy. While agriculture typically represents only a small portion of GDP in industrial countries, this is certainly not true in much of the developing world, where agriculture is often the dominant source of income and employment. A more complete representation of these economies is required to understand the likely impacts of agricultural trade reforms.

Economy-wide general equilibrium models provide a more complete representation of national economies. This requires the explicit specification of factor markets for land, labour and capital. In other words, the essential general equilibrium features are captured by including factor movements between sectors in addition to allowing for demand interactions. Economy-wide models capture implications of international trade for the economy as a whole, covering the circular flow of income and expenditure and accounting for interactions among different sectors of the economy.

At their core, computable general equilibrium (CGE) models (see Box 5) are concerned with resource allocation. This means tracking how the allocation of land, labour and capital responds to policy changes or exogenous developments. International trade is an arena where such effects can be an important outcome of policy choices. In the face of changing international prices, resources move between alternative uses within the domestic economy, or even between economies if production factors are internationally mobile.

CGE models attempt to measure the increase in economic welfare due to allocative efficiency improvement. Dynamic models attempt to measure the productivity gains that can arise from greater exposure to world markets, for example through economies of scale, improved technology and capital investment. Market imperfections such as partial price transmission, monopolistic market structures and similar frictions that abound in the agricultural markets of developing countries are, with the exception of imperfect competition, rarely included in CGE analyses.[9]

The main weakness of general equilibrium models is a direct consequence of their broader coverage. Because there is a trade-off between keeping the model workable and making it realistic enough to be useful to the policy community, CGE models are often constructed at fairly high levels of geographical and sector aggregation,[10] thus country- and commodity-specific detail

---

[9] For a recent example, see Roland-Holst (2004), who examines distance from market in Viet Nam and its impact on transmission of changes in prices at the international border.

[10] In recent years, the database compiled by the Global Trade Analysis Project (GTAP), a consortium involving international organizations such as FAO and World Bank, as well as governmental organizations and research institutes, has become the de facto standard for this type of analysis. All the studies considered here rely on this database. Some use the standard comparative-static perfectly competitive model provided by the GTAP consortium; others modify the model to include dynamic features and increasing returns to scale in non-agricultural sectors. More information is available at http://www.gtap.org.

> BOX 5
> **Key features of computable general equilibrium models**
>
> The main features of CGE models are summarized below.
> - Within each regional economy a standard CGE model covers inter-industry linkages through an input–output structure. Demand for factors of production is derived from cost minimization, given a sectoral production function that allows for substitution between inputs. Typically, substitution is allowed only between primary factors – land, labour, capital – while intermediate inputs are used in fixed proportion with output (Leontief technology).
> - The production structure is typically characterized as exhibiting constant returns to scale, and perfect competition is assumed to prevail in all markets. Each sector produces one homogeneous good that is perfectly substitutable domestically but substitutes imperfectly with foreign goods (Armington assumption). In addition to the distinction between domestic versus foreign goods, the multiregional nature of the model enables traded commodities to be differentiated according to their region of origin. In other words, bilateral trade flows are captured.
> - Factor markets for land, labour and capital are included in the model, and endowments for these primary factors are given and the factors are fully employed. Labour and capital are assumed to be fully mobile across domestic sectors, while land is imperfectly mobile and tied to agricultural production.
> - Consumer demand is derived from utility maximization under a budget constraint, and consumers allocate their expenditures over domestic and foreign goods. All factor markets and commodity markets are assumed to clear, which yields equilibrium solutions to factor and commodity prices as well as the corresponding equilibrium quantities.
> - Government policies are represented by various types of indirect taxes and subsidies, including import tariffs and export subsidies. In CGE models, policy measurement has converged on the concept of *ad valorem* price wedges, and all policy instruments are typically specified in this way
> - All regional economies are linked through bilateral commodity trade and through interregional investment flows. If a constant current account balance in all regions is assumed,

may be lost. Partial equilibrium models, in contrast, are often used to assess the commodity-specific impacts of reform. The two approaches are complementary, as each has its strengths and weaknesses.

Computable general equilibrium model results

Once the economy adjusts to the policy change, a new set of equilibrium conditions prevail. These new conditions are typically reported in terms of income or welfare effects, changes in trade flows and changes in returns to factors of production (e.g. wage rates). The sections below review the results from several CGE trade liberalization studies.

Welfare effects

Table 6 summarizes the welfare results of several recent CGE analyses of trade liberalization. The results of these studies are not completely comparable for a number of reasons. All use CGE models, but some use the standard GTAP model while others use customized models that allow for dynamic changes in productivity growth or departures from the standard assumption of perfect competition.

All of these studies rely on Version 5 of the GTAP database, except for Francois, van Meijl and van Tongeren (2005), which uses

then the difference between regional savings and investments is essentially predetermined; as a consequence, the aggregate level of the savings–investment balance is also predetermined. If endogenous determination of the current account balance is to be allowed for, the model must include a mechanism to redistribute aggregate savings over regions.
- Some models include a recursive sequence of temporary equilibria. Recursive models do generate time paths for endogenous variables, but there is no behavioural linkage among periods. As a result, the equilibrium solution in each period can be calculated without reference to earlier or later periods.
- Market imperfections are typically ignored in standard CGE models. Information problems, lack of infrastructure, monopolistic market structures and similar frictions abound in agricultural markets, especially in developing countries. However, CGE models rarely include those in the analysis. Only so-called "second-generation" models add increasing returns and imperfect competition in some of the sectors, allowing for estimates of scale and variety effects.
- The comparative-static analysis performed with CGE models does not reveal adjustment processes and possible adjustment costs involved when far-reaching policy changes are implemented. Policy-induced resource shifts will always entail income losses and adjustment processes for some people. The comparative-static CGE analysis typically sidesteps these issues and concentrates on the features of the new equilibrium in which the system settles after the policy change has been implemented.
- Relatively recent methodological developments have resulted in so-called "third-generation" models that include time-consistent forward-looking behaviour and endogenous savings rates, hence allowing for the modelling of short-run dynamics. While these models focus on savings–investment issues, including international capital flows, they could in principle be adapted to capture short- to medium-term real adjustment processes.

Source: Kehoe and Kehoe, 1994.

the newer Version 6. Version 6 differs in several important respects: it includes more countries and regions, is benchmarked to the year 2001 (instead of 1997) and uses more sophisticated measurement of levels of protection. Specifically, it includes existing preferential trade agreements and the conversion of specific tariffs to *ad valorem* equivalents. Therefore the new database captures the liberalization efforts that have been ongoing in the wake of the Uruguay Round as well as autonomous liberalization undertaken by many countries, especially in Asia after the Asian financial crisis of the late 1990s.

The studies reported in Table 6 look at different trade liberalization scenarios. Some concentrate on agricultural trade liberalization alone while others take a broader view and include non-agricultural market access, services and trade facilitation. Other studies assume that all barriers to agricultural trade are removed. These 100 percent liberalization scenarios assume that all forms of border protection, export subsidies and trade-distorting domestic support are eliminated. Others assume only a 50 percent cut in these trade barriers, while some focus only on tariffs, excluding other forms of support and protection.

Furthermore, the studies differ regarding which countries and regions liberalize. The most common scenarios in this regard are for global liberalization as opposed

TABLE 6
Welfare gains from CGE studies of trade liberalization

Study	Liberalization scenario	Notes	Welfare gains (billion $ 1997)				
			Global benefits from reforms			Benefits from agricultural reforms	
			All reforms	Non-agricultural reforms	Agricultural reforms	Developing countries	Developed countries
Anderson et al. (2001), GTAP	100 percent, all countries, all sectors, all policies		254	90	164	43	121
	Developing countries only		...	...	42	31	11
	Developed countries only		...	...	122	12	110
USDA (2001), CGE	100 percent, all countries, agriculture only, all policies	Static	...	...	31	3	28
		Dynamic	...	...	56	21	35
Francois, van Meijl and van Tongeren (2003), GTAP v5	100 percent, all countries, all sectors, all tariffs	Increasing returns to scale	366	257[1,2]	109	...	...
	50 percent, all countries, all sectors, all tariffs	Static	132	104[1,2]	28	11	17
		Dynamic	...	...	57	27	30
	50 percent, developing countries only	Static	...	...	11	6	5
		Dynamic	...	...	32	28	4
	50 percent, developed countries only	Static	...	...	17	5	12
		Dynamic	...	...	24	−0.7	25
Francois, van Meijl and van Tongeren (2005), GTAP v6[3]	50 percent, all countries, all sectors, all tariffs	Increasing returns to scale	168	138	30	7	24
	Developing countries only		...	...	10	10	0.5
	Developed countries only		...	...	20	−3	23
World Bank (2003)	100 percent, all countries, all sectors, all policies	Static	291	98	193	101	91
		Dynamic	518	156	358	240	117
	Developing countries only	Static	...	...	103	80	23
		Dynamic	...	...	185	167	19
	Developed countries only	Static	...	...	84	20	64
		Dynamic	...	...	174	75	100
IMF and World Bank (2002), GTAP	100 percent, all countries, agriculture only, all policies		...	...	128	30	98
	Developing countries only		...	...	27	22	5
	Developed countries only		...	...	102	9	93

[1] Includes services.
[2] Includes trade facilitation.
[3] Gains expressed in 2001 US dollar terms (billions).

to liberalization on the part of developed countries or developing countries only.

While these differences make direct comparisons of different model results problematic, the table nevertheless provides a useful overview of the range of potential welfare gains that may be possible from trade liberalization. Some general observations may be derived from these studies.

It should be noted that the income or welfare results from CGE models are typically expressed using a measure of economic welfare called "equivalent variation" (EV). The EV measures the change in income that would be equivalent to the proposed policy change – in other words, how much income should be given to (or taken away from) households to achieve the same welfare as the proposed policy change.[11]

The EV measures the potential change in welfare at the national level, but it does not consider distributive effects. Often, a policy change means that some people gain and others lose – rarely does everyone win. In fact, a positive EV means simply that the winners gain more than the losers lose. In economic terms, enough benefits will be generated by the policy change for the winners potentially to be able to compensate the losers.

The first three columns of Table 6 identify the model, the liberalization scenario, and the static or dynamic nature of the gains being reported. The remaining columns report the potential welfare gains arising from alternative liberalization scenarios. The fourth column reports the global welfare gains that are potentially available from liberalization in all sectors. The fifth and sixth columns show the potential gains from non-agricultural and agricultural liberalization, respectively. The final two columns report how the potential gains from agricultural liberalization would be distributed between developing countries and developed countries.

Comprehensive global trade liberalization
The two most comparable studies of comprehensive global trade liberalization are the first scenarios reported for Anderson *et al.* (2001) and the World Bank (2003). Both of these studies consider 100 percent elimination of all trade barriers in all sectors, using static models with standard economic assumptions. The potential global benefits in these studies range from $254 billion to $291 billion.[12]

Francois, van Meijl and van Tongeren (2003) found higher potential welfare gains of $366 billion in their comprehensive global liberalization scenario because their model allows for increasing returns to scale (firms are assumed to become more efficient as their size or scale of operation increases). The World Bank's (2003) dynamic model yields the highest potential welfare gains of all the major CGE trade studies to date, at $518 billion. The dynamic model goes beyond the simple static efficiency gains made possible by reallocating resources to more valuable activities. It supplements these efficiency gains with productivity gains that accrue when liberalization stimulates investment, for example in productivity-enhancing technology.

All the studies discussed so far assume comprehensive trade liberalization, i.e. 100 percent elimination of all trade barriers in all sectors by all countries. Francois, van Meijl and van Tongeren (2003; 2005) consider the potential welfare gains from less radical reforms. In these studies, trade barriers are reduced by only 50 percent. Not surprisingly, the potential gains are correspondingly smaller ($132 billion in their standard static model; $168 billion with increasing returns to scale).

Agriculture's contribution
Many CGE studies allow a comparison of the potential welfare gains arising from the liberalization of different sectors of the

---

[11] While the EV takes the new situation as a reference, an alternative measure known as "compensating variation" takes the old situation as the reference. It asks the hypothetical question: "What is the minimum amount of compensation after the price change in order to be as well off as before the change?"

[12] The World Bank (2005b) has revised its original study using the new Version 6 GTAP database. The global welfare impact from this revised model (not reported in Table 6) is $263 billion, slightly lower than their original estimate of $291 billion, reflecting among other things the liberalization that has occurred since the Version 5 database was released.

global economy or by different groups of countries. They also allow the gains to be disaggregated by region and country. The estimates of the potential benefits from agricultural liberalization differ markedly, depending on the model specification and the liberalization scenario, but some generalizations can be made.

Estimates of the potential static welfare gains from complete liberalization of the agriculture sector in the context of comprehensive reform range from $109 billion (Francois, van Meijl and van Tongeren, 2003) to $193 billion (World Bank, 2003). The USDA study (2001) found considerably smaller gains from reform of the agriculture sector ($31 billion in their static model). This study differs from the others in a number of key ways: it models agricultural liberalization only; it includes only WTO Members (China, which was not a Member at that time, was excluded) and it assumes that direct payments to farmers were completely decoupled from production.

The estimated gains in welfare from liberalizing all sectors range from one-third to two-thirds higher than from agricultural liberalization alone. In all the studies except those by Francois, van Meijl and van Tongeren (2003; 2005), agricultural reform yields a greater share of the overall gains than do non-agricultural reforms. The results obtained by Francois, van Meijl and van Tongeren can be explained by their more comprehensive treatment of non-agricultural reform (including services and trade facilitation) and their assumption of increasing returns to scale.

The largest share of estimated global income gains from agricultural liberalization accrues to industrial countries because these countries tend to have a higher incidence of economically inefficient agricultural policies in the first place, and they are the primary victims, economically speaking, of their own policies. Reduction, or even removal, of these distorting policy interventions leads to more economically efficient resource allocation, which is counted as a welfare gain.

Although the largest absolute gains (in US dollar terms) accrue to industrial countries, the largest relative gains in terms of GDP are consistently obtained by developing countries. Static welfare benefits for developing countries vary between $3 billion and $43 billion in the non-World Bank studies. This is equal to 0.2 percent and 0.7 percent of the GDP, respectively, of developing countries. In the World Bank study, welfare effects vary between $101 billion (static) and $240 billion (dynamic). The most optimistic World Bank scenario adds 1.7 percent to the GDP of developing countries.[13]

Even these larger GDP gains are fairly modest and are not sufficient to reduce significantly the average incidence of poverty in developing countries. This suggests that while trade liberalization benefits developing countries, liberalization alone will not enable them to achieve their Millennium Development Goals (MDGs) relating to poverty and hunger.

For developing countries, between 70 and 85 percent of the potential gains result from their own agricultural policy reforms. Lowering trade barriers among developing countries would open up increased opportunities for exports.

Finally, the model results show that some countries lose in the agricultural liberalization scenarios, even in the long run. Most countries gain but there are important exceptions. Net food-importing countries experience negative effects on their terms of trade as world food prices rise in the wake of policy changes. Current beneficiaries of preferential trade arrangements also lose as the value of the preferences are eroded. For these countries, the losses are not outweighed by efficiency gains from reallocating resources in agriculture alone. Such results highlight the importance of improved market access for non-agricultural exports from these countries.

Trade effects[14]

In addition to the welfare effects discussed above, another important dimension of the CGE modelling approach is the pattern of international trade. Indeed, some of these studies particularly stress the importance of tapping the potential for increased

---

[13] More recent unpublished estimates by the World Bank, in the context of ongoing work on trade and poverty, yield the same basic qualitative pattern of results. See, for example, Anderson and Martin (2005) and Hertel and Winters (2005).
[14] This section draws heavily on Francois, van Meijl and van Tongeren (2005).

TABLE 7
Bilateral trade: percentage change in value of bilateral import volumes

| | | Global trade round | | | | OECD-based trade round | | | |
| | | All commodities | | | | All commodities | | | |
From ↓	To →	EU-25	Developing countries	Other OECD	Total	EU-25	Developing countries	Other OECD	Total
EU-25		−2	17	10	4	−1	7	11	3
Developing countries		16	26	21	21	7	−2	8	5
Other OECD		12	22	6	12	11	9	7	8
Total		4	22	11	11	3	5	8	5
		Agriculture and food				Agriculture and food			
From ↓	To →	EU-25	Developing countries	Other OECD	Total	EU-25	Developing countries	Other OECD	Total
EU-25		−1	31	24	6	−1	3	12	1
Developing countries		25	44	24	32	17	5	16	12
Other OECD		31	36	25	29	27	14	22	21
Total		8	39	24	21	6	8	18	10

Source: Francois, van Meijl and van Tongeren, 2005.

South–South trade. Although volumes of trade among developing countries have displayed a remarkable rising trend in recent years, especially African–Asian trade, it is still the case that developing country exports are biased towards trade with the EU and the United States. Lowering trade barriers among developing countries would generate increased opportunities for South–South trade.

Francois, van Meijl and van Tongeren (2005) provides a good example of these results. Table 7 presents the estimated changes in bilateral trade flows for three groups of countries: the EU-25, developing countries and other OECD countries. Two scenarios are considered: a global trade round scenario, in which all countries liberalize all sectors, and an OECD-based scenario, in which only OECD countries engage in reforms. Finally, results are shown for trade in all goods and trade in agriculture and food.

The upper left panel of Table 7 shows changes in total trade flows under the global trade round scenario. Global trade expands by 11 percent while intra-EU trade declines by 2 percent. As a consequence of diminishing intra-EU trade preferences, suppliers from developing countries expand their exports to the EU by 16 percent, and realize the most impressive growth in market share on European markets.

Developing countries obtain the highest overall growth in exports (21 percent). They are stimulated to expand exports to all destinations, but the greatest surge is observed in trade among developing countries themselves.

In the lower-left part of the table agricultural trade is presented separately from the aggregate. By comparing these numbers with those for all goods it can be seen that developing country exports are mainly driven by agricultural exports. Developing country agricultural exports expand by 32 percent, with by far the largest growth occurring in trade among developing countries.

Other OECD countries also see strong growth in agricultural exports, especially to the EU and the developing countries. This group comprises Australia, New Zealand and the United States, which are themselves important agricultural exporters.

Turning to the right-hand panel of Table 7, an OECD-based round, with developing countries not participating in reform, trade growth is reduced for both country groups but especially for developing countries. Intra-developing country trade shrinks relative to the base. This points to yet more

TABLE 8
Effects of trade liberalization on unskilled wages by sector and scenario (percentage change)

	Total		Global trade round				OECD-based trade round			
	Global reform	OECD-based reform	Agriculture	Manufac.[1] tariffs	Service	Trade facilitation	Agriculture	Manufac. tariffs	Service	Trade facilitation
Europe										
France	1.3	1.2	0.4	0.3	0.5	0.1	0.5	0.1	0.5	0.1
Germany	1.3	1.1	0.4	0.5	0.3	0.2	0.4	0.3	0.3	0.1
Netherlands	1.3	1.1	0.5	0.5	−0.1	0.4	0.5	0.4	−0.1	0.3
Rest of EU-15	0.9	0.7	0.4	0.3	0.0	0.2	0.4	0.2	0.0	0.1
EU-10	0.3	0.3	0.3	−0.1	0.1	0.1	0.3	−0.1	0.1	0.1
Africa and the Near East										
Mediterranean region	1.6	0.0	0.4	0.7	0.1	0.4	−0.1	0.0	0.1	0.1
South Africa	2.0	0.7	0.0	0.7	0.7	0.5	−0.2	0.1	0.6	0.1
Sub-Saharan Africa	3.1	0.9	0.8	0.8	1.0	0.5	0.0	−0.1	0.9	0.2
Americas										
North America	0.1	0.1	0.0	0.1	0.0	0.1	0.0	0.0	0.0	0.1
South America	0.4	−0.1	−0.1	0.2	0.2	0.2	−0.3	0.1	0.1	0.0
Asia and the Pacific										
China	−0.3	−0.6	0.1	−0.7	0.2	0.2	−0.2	−0.5	0.1	0.0
India	3.1	0.2	0.9	1.2	0.8	0.3	−0.1	0.2	0.0	0.1
High-income Asia	1.6	1.3	0.7	0.4	0.1	0.3	0.7	0.3	0.1	0.3
Other Asian and Pacific countries	4.5	0.5	0.4	2.4	0.0	1.7	−0.2	0.3	0.0	0.5
Australia and New Zealand	1.3	1.1	0.1	0.5	0.4	0.3	0.1	0.4	0.4	0.3
Rest of world	0.2	0.1	−0.3	0.2	0.1	0.2	−0.3	0.0	0.1	0.2

[1] Manufactures.
Source: Based on simulation results from Francois, van Meijl and van Tongeren, 2003.

trade diversion effects in the face of OECD countries lowering their trade barriers while non-OECD barriers remain in place.

In the OECD-based scenario, developing country exports to developed economies expand at a slower pace than in the broader liberalization scenario. This is because failure to engage in their own reforms precludes specialization, and insufficient resources are freed to allow expansion in export-oriented industries. The slower export growth implies that insufficient foreign exchange is earned to finance an expansion in imports.[15]

---

[15] A technical term in trade theory, Lerner symmetry, is relevant here. Import barriers ultimately suppress exports. This is very evident in the pattern of developing country exports.

Wage effects

Table 8 reports the impact of trade policy reform on unskilled wages, one of the key avenues through which trade influences poverty. This CGE study by Francois, van Meijl and van Tongeren (2003) considers a 50 percent reduction in domestic support, export subsidies and import protection in agriculture as well as manufacturing and services. This approach allows the broader impact of trade policy on incomes (and hence on income-related aspects of food security) to be gauged.

In general, for the middle- and low-income groupings shown, agriculture is far more important for unskilled labour earnings in developing countries than it is in the OECD countries. At the same time,

TABLE 9
Impacts of policy reform on world commodity prices

	Partial policy reform (phase-out of market price support)	Complete policy reform (phase-out of all support)
	(Change in real prices relative to baseline[1])	
Cereals	103	111
Wheat	104	119
Rice	104	111
Maize	99	106
Milk and dairy products	111	117
Beef	106	108
Sheep and goat meat	104	105
Pig meat	102	103
Poultry meat	103	104

[1] Baseline = 100.
Source: FAO, 2003a.

though, it is liberalization outside the OECD countries – primarily own-policy reform – that leads to the bulk of agriculture-related wage gains for developing countries. What really matters will vary for different countries and regions. Hence, for North Africa and the Near East, unskilled workers stand to gain the most from agricultural policy reform at home. The same is also true in South Africa. In India, on the other hand, manufacturing liberalization (such as clothing tariffs in middle-income countries) is at least as important as agriculture. The same holds for the group of other Asian and Pacific countries.

The wage results in CGE studies provide a bridge to the household impact of agricultural trade, the topic of the next chapter. It should be noted that unskilled workers are not necessarily agricultural workers; in fact, unskilled wages often underpin the income of urban households in low- and lower-middle-income countries. As such, rising unskilled labour earnings in urban households may go hand-in-hand with falling earnings in rural households. Unravelling this mix of rural and urban households within CGE studies requires a move to models that include household data.

One conclusion that can be drawn at this stage is that agricultural trade can have a significant impact on household wage earnings. However, non-agricultural trade can have an equal or even greater impact on wage earnings. To the extent that food security is a function of incomes and the ability to secure sufficient food through monetary means, then food security depends on trade outside agriculture as well as on trade in agricultural products.

Agriculture sector model results

As mentioned above, partial equilibrium agriculture sector models can provide a more detailed picture of the effects of trade liberalization on individual commodity markets. FAO conducted a detailed assessment of the potential commodity-level impacts of agricultural reform, taking as its baseline a consistent set of long-run production and trade projections. This model takes into consideration the potential of countries to respond to policy changes for different types of commodity (FAO, 2003a).

Table 9 reports the results of two liberalization scenarios. In the first, all market price support to agriculture is phased out. In the second, all support and protection to agriculture is phased out in developed and developing countries. As in the CGE analyses discussed above, the majority of the benefits under this scenario accrue to the OECD countries in the form of lower consumer prices for previously protected products.

Even in the more complete policy reform scenario, the price impacts are likely to be modest. The most significant commodity market changes are expected to occur for temperate-zone commodities such as cereals, meat and dairy products that account for the majority of OECD policy distortions. Production of these commodities would fall in the currently subsidizing countries and expand in the non-subsidizing countries, primarily other OECD producers such as Australia and New Zealand and some developing countries where temperate zone commodities are produced competitively, such as Argentina, Brazil and Thailand.

Products such as rice and sugar, which are highly subsidized and for which many developing countries are competitive producers, could yield particularly large gains for developing countries. On balance, world prices for temperate zone commodities would increase by 5–20 percent, depending on the initial level of market distortion and the capacity of other producers to expand output. These price effects are relatively small because considerable production potential exists for most commodities and because the simultaneous liberalization of all commodities would tend to have offsetting effects.

Developing countries would gain relatively little from further liberalization of tropical commodities such as coffee and cocoa because import barriers in the OECD countries are already fairly low and consumption is saturated. There is some potential for gains for these commodities in other developing countries, where import barriers are relatively high. The ability of farmers in developing countries to benefit from liberalization depends to a great extent on the domestic agricultural policies of their own countries, which often place them at a disadvantage through high effective rates of taxation, poor infrastructure and inefficient marketing systems (FAO, 2003a).

Key findings

The economic benefits that could result from comprehensive reform of agricultural policies are potentially important, particularly when combined with reforms in other sectors. But the reform process will necessarily involve adjustment costs. Policy-makers need to understand the likely impacts of agricultural trade policy reforms before they are agreed, so that proper policies can be put in place to assist in the adjustment process.

While the various economic models used to assess the potential impacts of trade liberalization differ in their details, a number of general observations are fairly consistent across a wide variety of models and practitioners.

- Trade liberalization in agriculture is likely to generate positive economic benefits at the global level and for most – but not all – participating countries. Liberalization scenarios that involve all sectors and all regions tend to generate higher benefits than scenarios where some sectors or regions are excluded.
- The largest absolute gains from agricultural liberalization accrue to the developed countries where agricultural markets are most distorted. These gains go primarily to consumers in OECD countries where import barriers are currently high and to exporters in other OECD countries.
- Developing countries gain more as a share of current GDP because agriculture is much more important in their economies. Some developing country exporters of temperate zone agricultural products gain from OECD liberalization, but the biggest gains for developing countries derive from liberalization among themselves. Virtually all of the growth in agricultural markets over the next 30 years will occur in developing countries, so the potential gains from further opening these markets are substantial.
- Some developing countries, particularly NFIDCs and countries currently receiving preferential access to OECD markets, lose from the OECD liberalization, even in the long run. The special circumstances of these countries must be taken into consideration.
- The potential global welfare gains from trade liberalization are, on the whole, relatively modest compared with global GDP. Dynamic gains are worth about twice as much as static gains alone, and developing countries stand to gain proportionately more from these dynamic gains. Countries should pursue aggressive pro-poor growth strategies to take advantage of these potential dynamic gains.

- Trade liberalization in agriculture and other sectors could contribute significantly to raising the wages of unskilled and low-skilled workers in developing countries, who are often among the poorest of the poor. As the next chapter illustrates, labour markets are one of the most important avenues through which trade liberalization affects poverty at the household level. The ability of poor people in developing countries to take advantage of the opportunities generated by trade reform depends crucially on the policy environment created by their own governments.

5. Poverty impacts of agricultural trade reforms

The impact of trade policy on poverty, food security and inequality in developing countries is at the centre of a crowded international debate on the role of international trade in development. The current Doha Round of trade negotiations makes development and poverty impacts a top priority. In addition, the Millennium Declaration underscores the importance of international trade in the context of development and the elimination of poverty. In the Millennium Declaration, governments committed themselves, *inter alia*, to an open, equitable, rule-based, predictable and non-discriminatory multilateral trading system.

Developing countries place great emphasis on assessing the distributional and food security consequences of trade liberalization and their domestic policy reform efforts. This growing interest has fuelled a wealth of empirical studies on the links between trade policy and complementary domestic policies and their impacts on inequality and poverty.

This chapter reviews much of this empirical evidence and examines the impacts of both unilateral domestic agricultural policy and trade reforms and multilateral trade liberalization on poverty.[16] Attempts to correlate trade and trade liberalization positively with economic growth have a divisive and ambiguous history (Rodríguez and Rodrik, 1999). Studies establishing positive links between economic growth and poverty reduction are more convincing (see Bardhan, 2004, for a recent review).

Emphasis is given to agricultural trade policies. However, trade liberalization is generally an economy-wide phenomenon, with tariff cuts occurring across a wide range of commodities, so the review is not restricted to episodes where only agricultural trade is liberalized. Furthermore, given the difficulty of isolating the effects of trade policies, the impact of other types of external shock that alter relative prices of tradeable and non-tradeable goods is considered.

By examining the ways in which households adjust to such external shocks, a great deal can be learned about how they would respond to sharp reductions in tariffs, or significant changes in a country's international terms of trade engendered by trade liberalization.

Food-insecure and poor households in developing countries are very diverse, and they are affected in different ways by agricultural trade reforms. While this discussion focuses most of its attention on how rural households respond to various trade reforms, to understand the impact of a given trade reform on national food security and poverty, the effect on urban households is equally important.

Agriculture's role in poverty reduction

The economic linkages among agriculture, trade and poverty are complex. Agriculture plays a central role in the lives of the poor, both as the main source of their livelihoods and their main consumption expenditure. Thus, to the extent that agriculture is affected by trade, trade has implications for poverty and food security.

Poverty is multidimensional and dynamic, with large numbers of vulnerable families moving in and out of poverty over time. Poverty means high levels of deprivation, vulnerability to risk and powerlessness. Seeking a better understanding of the links among poverty, economic growth, income distribution and trade remain a permanent issue in development literature (Box 6).

Agricultural growth is particularly important for poverty reduction and food

---

[16] The conceptual framework for the trade–poverty linkages presented by Winters (2002), and the literature reviews offered by Winters, McCulloch and McKay (2004) and Hertel and Reimer (2004), provide the core background materials for this chapter

> **BOX 6**
> **What do we know about poverty reduction?**
>
> Important lessons for poverty reduction include the following:
> - Poverty cannot be reduced without economic growth (or mean incomes rising) and the economic growth is neutral to income distribution or reduces income inequality.
> - Large income inequalities are bad for poverty reduction and economic growth.
> - Public investment and incentives for better nutrition, health and education benefit the poor through increased consumption and through higher future incomes.
> - Capital-intensive technologies, import substitution and urban bias growth processes induced by price, trade and public expenditure policies are not good for reducing poverty.
> - Agricultural growth, with low asset concentration and labour-intensive technologies, is good for poverty reduction.
>
> ---
>
> *Sources:* FAO, 1993; Atkinson and Bourguignon, 2000; Lipton and Ravallion, 1995; Bruno, Ravallion and Squire, 1998; Ravallion and Datt, 1999; Aghion, Caroli and Garica-Penalosa, 1999; Khan, 2003.

security in developing countries. A number of factors help explain why.

Poverty as a rural phenomenon

First, poverty in developing countries is concentrated in rural areas, especially in those countries where the levels of undernourishment are greater than 25 percent. Most estimates suggest that more than two-thirds of the poor live in rural areas (FAO, 2004b).

While demographic and migration trends are shifting the poverty balance towards urban areas, the majority of the poor will continue to live in the countryside for at least a few more decades. In general, the more remote the location the greater is the incidence of poverty.

Moreover, urban poverty is to a large extent the result of rural deprivation, which encourages rural–urban migration. No sustainable reduction in poverty and undernourishment is possible without development of the rural areas.

Country-level surveys highlight the disparity between rural and urban areas. For example, the percentage difference between rural poverty and urban poverty in seven countries (as reported in their World Bank Poverty Reduction Strategy Papers [PRSPs]), ranged from 9 percent in Mozambique to 35 percent in Burkina-Faso, 38 percent in Nicaragua, 41 percent in Mauritania and 42 percent in Bolivia (Ingco and Nash,

2004). Furthermore, it is not just the poverty indicators that highlight the rural–urban disparity: rural populations score consistently lower on every quality of life indicator.

Economic importance of agriculture

Second, the central role for agriculture in supporting poverty reduction and food security is underlined by the relative economic importance of the sector for developing countries. Seemingly paradoxically, agriculture represents a larger share of the economy in those countries with the highest percentage of poor and undernourished people in their populations.

Figure 15 presents the percentage share of agriculture in total GDP for developing countries grouped according to the prevalence of undernourishment. For countries where more than one-third of the population are undernourished, the share is almost 25 percent; this share declines with decreasing levels of undernourishment in the population.

Agriculture and employment

Third, most of the income-earning opportunities for the rural poor are related directly or indirectly to agriculture (Figure 16). For developing countries as a whole, agriculture accounts for about 55 percent of employment. Again, the share of agricultural employment in total employment is higher for countries with a

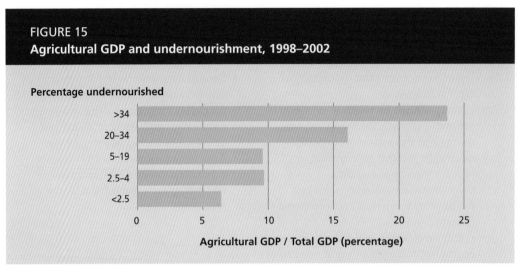

Source: FAO and World Bank.

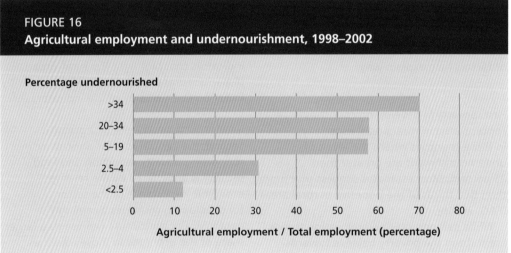

Source: FAO and World Bank.

higher prevalence of undernourishment and reaches as much as 70 percent, on average, for the countries where 34 percent or more of the population are undernourished.

The rural poor face a diverse set of problems, with an equally diverse set of solutions. Many of the solutions, however, are linked to an expanding agriculture sector where the poor can find jobs related to producing, supplying, storing, transporting, processing and reselling inputs, services and products.

Higher producer incomes, more jobs and higher wages for labourers lead to increased demand for goods and services that are often difficult to trade over long distances. Additional job opportunities emerge in non-farm activities to meet increased demand for basic non-farm products and services – including tools, blacksmithing, carpentry, clothes and locally processed foods, to name a few. These and related goods and services tend to be produced and provided locally, with labour-intensive methods, and so have great potential to create employment and alleviate poverty. Surveys in four African countries suggest that between one-third and two-thirds of income growth in rural areas is spent on such local goods and services (FAO, 2003a).

Agriculture and pro-poor growth
The concentration of poverty in rural areas and the importance of the agriculture sector in output and employment among the poor all point to a central role for the sector in addressing poverty.

Such agriculture-led growth often lowers poverty in both urban and rural areas.

A major study by FAO examined the roles of agriculture in 11 developing countries, concluding that the pro-poor role of agriculture can be dramatic and much more effective in reducing poverty and hunger than other sectors in both rural and urban areas (FAO, 2004c).

In each country case study, researchers analysed the extent to which agricultural growth reduced poverty (i.e. the elasticities of national poverty levels with respect to agricultural growth). In some countries, the studies also assessed agriculture's contribution to poverty reduction relative to other sectors and in rural areas.

This component of the FAO study, known as the Roles of Agriculture Research Project (ROA), drew its inspiration from a 1996 study by Ravallion and Datt in which they compared the poverty reduction effects of agricultural growth with those of industry and services in India. The authors of the ROA study found national-level poverty elasticities with respect to agricultural growth ranging from –1.2 to –1.9. The urban poverty elasticities ranged from –0.4 to –0.5.

The study also explored how poverty is being reduced. Four channels for poverty reduction were considered: falling real food prices, creating employment, higher real wages and rising incomes for small farm households.

The results demonstrate that agricultural growth has a strong and positive impact on poverty reduction, often significantly greater than that of other economic sectors. Noticeably, this pro-poor outcome was observed not only for the poorest and most agrarian countries (Ethiopia and Mali), but also for the higher-income economies (Chile and Mexico).

The results also suggest that poverty reduction policies should take into consideration the strategic importance of agricultural growth and its transformation, the output mix (especially towards labour-intensive exports) and the various channels through which agriculture may contribute to poverty alleviation (Valdés and Foster, 2003).

Finally, agriculture's evolving economic linkages provide multiple opportunities to contribute to growth, poverty reduction and food security (Vogel, 1994; Timmer 1995; Anderson, 2002; FAO, 2003a; Sarris 2003; de Ferranti et al., 2005).

In agrarian societies with few trading opportunities, most resources are devoted to the provision of food. As national incomes rise, the demand for food increases much more slowly than other goods and services. New technologies for agriculture lead to expanding food supplies per hectare and per worker and the increasingly modernizing economies use more intermediate inputs purchased from other sectors.

Agriculture's share in total GDP declines with economic growth as post-farmgate activities are taken over by specialists in the service sector and become more commercialized. Commercial development occurs on the input side also, as producers substitute chemicals and machines for labour.

Although agriculture's share of GDP may fall relative to industry and services, the sector can nevertheless grow in absolute terms, evolving increasingly complex linkages with non-agricultural sectors. Agriculture's productive and institutional links with the rest of the economy produce demand incentives (rural household consumer demand) and supply incentives (agricultural goods without rising prices) that promote modernization.

While poverty reduction channels are not unique to agriculture, the pro-poor role of agricultural growth raises several important questions: Is agriculture receiving the priority it deserves in national policy-making? What role can trade play in making the most of the sector's potential? What types of domestic policies and public investments are needed to make agricultural trade work for the poor and food-insecure?

Trade's role in poverty reduction

FAO has long argued the virtues of trade's contributions to economic growth and resource efficiency, as well as its contributions to food security by providing a stable source of lower priced food from abroad. In addition, from a trade perspective, agriculture is particularly important for countries with a high prevalence of undernourishment (Figure 17).

For instance, for developing countries as a whole, agricultural products (including fisheries and forestry) account for about 9 percent of total trade (exports plus

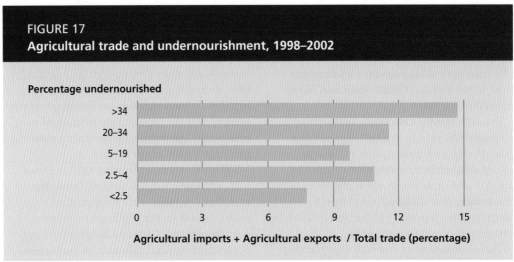

FIGURE 17
Agricultural trade and undernourishment, 1998–2002

Source: FAO and World Bank.

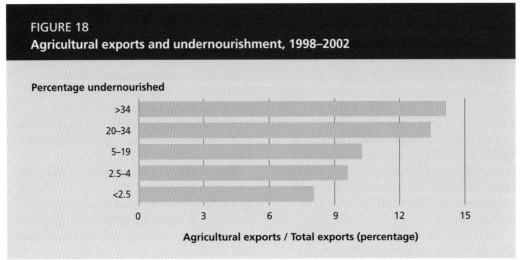

FIGURE 18
Agricultural exports and undernourishment, 1998–2002

Source: FAO and World Bank.

imports), while for the countries with the highest prevalence of undernourished, the share is almost 15 percent. These numbers reflect an economy with lower levels of industrialization and little diversification within their agriculture sectors.

Looking at exports only, the country group with the highest levels of undernourishment is the most heavily dependent on agriculture, which accounts for more than 14 percent of their total exports (Figure 18). In spite of their high dependence on agriculture for income, employment and export earnings, countries in this group nevertheless spend more than 15 percent of their total import budget, and on average more than 12 percent of their total export earnings, to finance food imports (Figures 19 and 20).

Although the share of agricultural trade in total trade is high for those countries with the worst levels of undernourishment, their agriculture sectors are relatively less integrated into international markets. This is illustrated by Figure 21, which presents the ratio of agricultural trade to agricultural GDP for country groups by level of undernourishment in the population.

Trade–poverty linkages

Trade–poverty linkages are complex and diverse. The first linkage is at the border. When a country liberalizes its own trade policy by, for example, reducing import tariffs, this results in lower prices for imported goods at the border. When other countries liberalize their trade policies, this

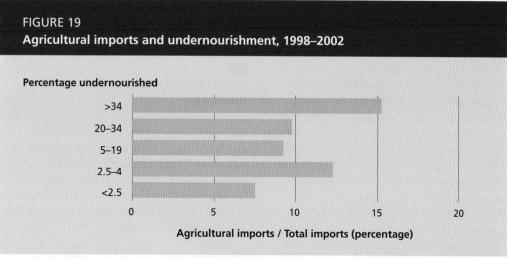

FIGURE 19
Agricultural imports and undernourishment, 1998–2002

Source: FAO and World Bank.

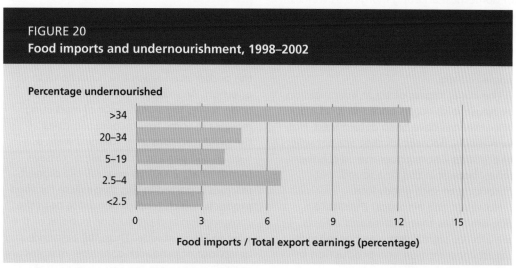

FIGURE 20
Food imports and undernourishment, 1998–2002

Source: FAO and World Bank.

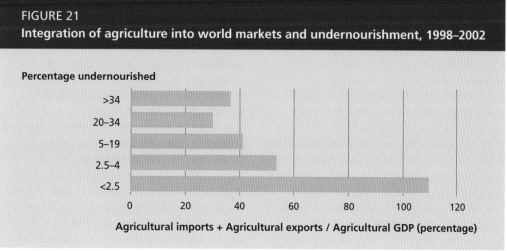

FIGURE 21
Integration of agriculture into world markets and undernourishment, 1998–2002

Source: FAO and World Bank.

affects the border prices of goods imported and exported by the first country. The direction and magnitude of the initial border price changes depend on the precise policy reforms being undertaken. As discussed in Chapter 4, the elimination of all forms of support and protection to agriculture by the OECD countries would be expected to increase the border prices of temperate-zone agricultural products by about 5–20 percent.

From the border, the focus moves to how prices are transmitted to producers and consumers, and to households in general. The extent to which households and businesses in the economy experience these price changes varies, and depends on the quality of infrastructure and the behaviour of domestic marketing margins as well as geographical factors. The empirical literature confirms this, sometimes wide, variance in the degree of price transmission from the border to the local market, even within a single country.

The initial impact of trade liberalization on households occurs once the local market price changes have been determined. Not surprisingly, households that are net sellers of products whose prices rise, in relative terms, benefit in this first round. Net purchasers of such goods lose.

However, the empirical literature demonstrates that first-round effects are altered significantly as households adjust consumption and production in response to changing relative prices. In this second round of effects, households modify their consumption basket, adjust working hours and possibly change their occupation. Evidence also suggests that changes in relative prices can even affect a household's long-term investment in human capital.

As households change their spending levels and employment patterns and as landowners and firms adjust their hiring, a wide range of effects ripple throughout the economy. For example, trade reforms that stimulate agricultural production often lead to a general increase in unskilled wages. This, in turn, benefits households that are net suppliers of unskilled labour. Finally, the long-run growth effects associated with trade liberalization need to be considered, including increases in firm productivity due to access to new inputs and technologies as well as potential gains arising from the disciplining effect of foreign competition on domestic profit margins.

Agricultural trade reform and poverty

The importance of the agriculture sector and trade for poverty reduction are well established. Less well understood are the mechanisms through which trade liberalization in agriculture affects the poor and the capacity of the poor to adjust to the new policy environment.

Transmission of prices to consumers and producers

One of the more important issues to address when considering the potential impact of trade reforms on the poor is the extent to which changes in prices at the border even reach the households in question. An example from Mozambique underscores the significance of marketing margins in some low-income countries: the producer–consumer margins were as high as 300 percent in the case of cassava (Arndt et al., 2000). In general, the biggest margins reported in this study were for food products, which tend to dominate both the consumption and production bundles of the poor in Mozambique. Thus, the existence and behaviour of producer–consumer margins are critically important for any poverty study.

If these marketing costs are solely a function of the quantity transported (i.e. *specific* as opposed to *ad valorem* in nature), then they dampen the impact of world commodity price changes on domestic consumers and at the same time exaggerate the impact of such price changes on producers of export products (Winters, McCulloch and McKay, 2004).

In Uganda, for example, transport margins protected domestic sales while taxing taxed exports over the decade 1987–97 (Milner, Morrissey and Rudaheranwa, 2001). Uganda's traditional exports include coffee, tea, cotton and tobacco and while a series of trade policy reforms over this period largely eliminated the implicit taxation of exports through trade policies, the implicit taxation caused by poor infrastructure and high transport costs remained very high

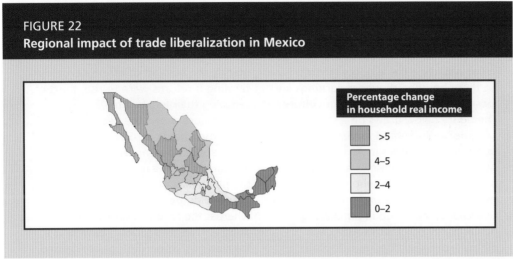

FIGURE 22
Regional impact of trade liberalization in Mexico

Source: Nicita, 2004.

relative to that of competitor countries such as Kenya. The transport-induced effective rate of taxation on exports from Uganda in 1994 was estimated to be equal to nearly two-thirds of value-added. Effective protection for domestic sales provided by the transport-induced trade barriers remained high throughout this period of reform. These "non-policy" barriers to trade represent one important reason for the sluggish response of the Ugandan economy to the extensive trade policy reforms undertaken over this period.

In Viet Nam, the geographical fragmentation of markets is a critical issue. There is a direct correlation between access to large markets and the transmission of border price changes to internal markets. For many isolated economic regions in the country, international trade (and even economic activities in other regions) is largely irrelevant (Roland-Holst, 2004).

Another recent study analysed the impact of NAFTA on rural producers and consumers in Mexico, addressing the question of price transmission from the border to domestic markets (Nicita, 2004). This report incorporates differential pass-through of Mexican tariff changes by region – estimated to be a function of the region's distance from the United States, the primary source of most Mexican imports.

Consistent with other studies of this nature, Nicita found incomplete pass-through of the tariff changes to consumers in Mexico, with the extent of pass-through being smaller for agricultural commodities than for manufactured goods. When coupled with a rapid erosion of pass-through with increasing distance from the border, this means reductions in agricultural tariffs have little or no impact on the more remote regions of Mexico. High transportation costs and the greater competition from domestic sources faced by these products are the reasons for the low pass-through for agricultural products. Therefore, local production quickly becomes more profitable as one moves away from the border.

Figure 22 reports Nicita's estimates of the regional welfare impacts of trade reforms undertaken by Mexico in the 1990s. The study illustrates a considerable regional variation in impact, with households in some regions gaining more than 5 percent of real income, while others register negligible gains. Trade liberalization can also have an impact on marketing margins, particularly to the extent that it opens up the opportunity for investment in logistics, transport and marketing activities that may have previously been dominated by monopolies. Badiane and Kherallah (1999) also explore this aspect with reference to several African countries.

Initial impacts of price changes on households

For self-employed, rural producers, the impact of a given set of border price changes, transmitted to the "farmgate" depends largely on their net sales position. Box 7 explores the impact of trade reforms on those households whose earnings are most dependent on agriculture.

BOX 7
Agricultural households

What impact do trade policy reforms have on those households whose earnings are most directly dependent on agriculture? The figure below draws on a set of 14 national household surveys for a selection of countries in Africa, Latin America and Southeast Asia. The figure plots the share of households that are specialized in agricultural income against GDP/capita, measured in purchasing power parity (PPP) terms. Here, we define "specialization" as referring to households that earn 95 percent or more of their income from agricultural profits. So, not only do they work full-time in agriculture, but they are also self-employed. This means that it may be difficult to switch to other activities if returns from farming were to fall. Likewise, because they are fully employed in agriculture, they are unable to increase quickly the amount of effort devoted to farming if returns were to rise, short of reducing their leisure time.

The figure shows the negative correlation between GDP per capita and the share of households specialized in agriculture. In the poorest country in the sample, Malawi, nearly 40 percent of households are specialized in farming, whereas the richest countries in the sample, Chile and Mexico, have only a fraction of that percentage specialized in agriculture. Of course, there are some outliers. For example, Viet Nam is a low-income country that also appears to have a low level of agricultural specialization. However, it is clear that, for many developing countries, the agriculture-specialized segment of the population is important, and this is generally inversely related to per capita GDP.

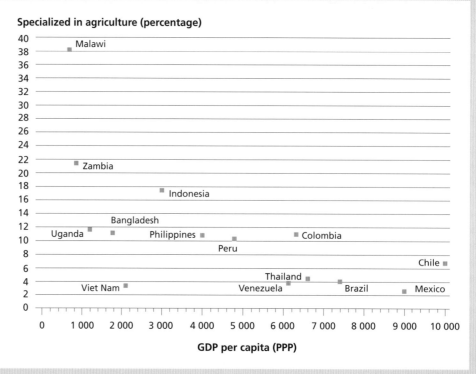

The share of agriculture-specialized households declines with per capita GDP

Source: Hertel et al., 2004.

FIGURE 23
Initial impact of WTO accession on rural and urban household real income in China

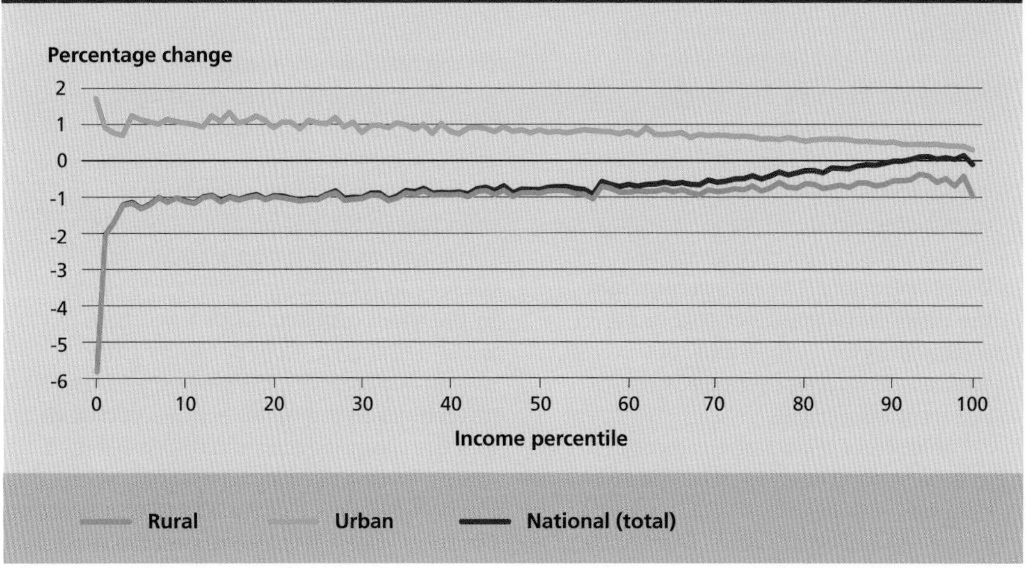

Percentage change (y-axis: -6 to 2) vs **Income percentile** (x-axis: 0 to 100)

Legend: Rural — Urban — National (total)

Source: Chen and Ravallion, 2003.

If the household is a net exporter of a product whose price has risen, it benefits; if it is a net importer, then it loses. Summing over the net sales-weighted price changes estimates the overall change in household welfare. This approach was used to assess the *ex-ante* household welfare impacts of trade liberalization in the cases of China's WTO accession (Chen and Ravallion, 2003) and Morocco's unilateral trade liberalization (Ravallion and Lokshin, 2004).[17]

The China study found that the initial trade reform impact harms rural households, while benefiting urban households. This is because China is required to reduce protection on a number of important agricultural imports, whereas the average rate of manufacturing protection is quite low for most sectors as a result of the widespread use of duty drawbacks for manufactured goods and generally lower average tariffs.

The largest percentage change in welfare is for the poorest households (Figure 23), with the poorest rural households losing more than 2 percent of their income and the poorest urban households gaining nearly 2 percent of initial income. Overall, however, the effects of WTO accession on China appear to be rather modest, partly because the deepest tariff cuts had already been made in anticipation of this agreement, but also because of the difficulty of quantifying the potential price effects of the accession agreement as it pertains to foreign commercial presence in the services sector of China (Walmsley, Hertel and Ianchovichina, 2005).

Tariff cuts on cereal imports in Morocco have adverse impacts on rural poverty while contributing to a fall in urban poverty. One of the more interesting results in the Morocco study is the decomposition of the aggregate change in inequality (which increases) into its vertical and horizontal components. The vertical component evaluates the change in inequality arising from differential impacts on households at different pre-reform levels of welfare. By this measure, inequality declines slightly following reforms because the poor tend to spend a disproportionate share of their income on grains, and grain prices fall under the reforms.

However, the dominant impact of reforms is to increase horizontal inequality – which is measured by assessing the impact on different households at the same level of pre-reform welfare. This is explained by the fact that many of the rural poor in Morocco tend to be net sellers of grains, and thereby

---

[17] However, like most studies of this sort, these two do not take account of incomplete price transmission from the border to the local level.

> **BOX 8**
> **Impact of agricultural liberalization on poverty in Brazil**
>
> Because different households have different income profiles, they are affected differently by changes in policy. To illustrate this point, Hertel and Ivanic (2005) use a global general equilibrium model to track the impact of a global round of agricultural trade liberalization on the different income strata of Brazilian society. The results highlight the differential impact that changes in consumer prices, urban and rural wages, and capital income can have across different households.
>
> The poverty impact across income strata in Brazil is illustrated in the table below. Basically, with poverty rising in some strata and falling in others, it is not clear, *a priori*, whether overall poverty in Brazil will rise or fall following multilateral agricultural trade liberalization. Focusing on the relative concentration of poverty in these strata does shed some light on the question, however. The poverty rate among the agriculture-specialized households in Brazil is much higher than that in the nation as a whole. As a consequence, this group accounts for 27.5 percent of total poverty – roughly equal to the share contributed by the urban, wage-earning stratum. Because of the overall importance of self-employed farm households in the national poverty picture, and the sharp reduction in their poverty rate following agricultural liberalization, the national poverty rate also falls in both the short-run (–2.9 percent) and the long-run (–1.6 percent) despite the increases in poverty in other strata.
>
> **Agricultural trade liberalization and poverty: impacts in Brazil**
>
> | Stratum | Initial poverty share | Percentage change in poverty | |
> |---|---|---|---|
> | | | Short-run | Long-run |
> | Agricultural | 0.275 | –11.5 | –1.9 |
> | Non-agricultural | 0.111 | 1.3 | –1 |
> | Urban labour | 0.276 | 0.8 | –2.2 |
> | Rural labour | 0.154 | 0.5 | –1.3 |
> | Urban diverse | 0.039 | –0.8 | –2.1 |
> | Rural diverse | 0.039 | –4.5 | –1.7 |
> | Total | | –2.9 | –1.6 |
>
> *Source:* Hertel and Ivanic, 2005.

lose from the price declines; the poor in urban areas are net buyers and therefore gain. Because the horizontal component dominates, overall inequality rises following cereals import reforms in Morocco.

Box 8 presents the impact of agricultural liberalization for households with different income profiles in Brazil, where households specializing in agriculture account for more than one-quarter of total poverty.

A study of the distributional consequences of devaluation in Rwanda emphasizes the importance of home production (Minot, 1998). This study concluded that a devaluation that raises the price of tradeables relative to non-tradeables by about 40 percent has only a modest negative impact on the poorest rural households, whose cash purchases comprise only about one-third of total expenditure.

The largest proportional losses accrue to the wealthiest urban households, who devote 96 percent of their income to cash purchases. Because one of the most important features of trade liberalization is often a change in the real exchange rate, this point is worth bearing in mind. Rural and low-income households are likely to be less severely affected either positively or negatively, because home production is more prominent in their overall consumption profile.

How households adjust to terms of trade shocks

With the exception of the Rwanda study, the analyses referred to in the preceding sections have simply used the households' initial sales and expenditure weights in the welfare analysis, thereby ignoring any potential for adjustment in response to the price changes. Of course, households tend to reduce consumption of higher-priced goods, while at the same time increasing their supply, thereby enhancing the potential for gains from a given set of exogenous price changes. Some studies have attempted to measure the potential for such adjustments and how they can affect the impact of external shocks on the rural poor.

One recent study of the potential of consumer substitution in the face of higher border prices estimated the effect of the Indonesian financial crisis on consumer welfare assuming (i) no substitution (as with the studies by Ravallion and co-authors) and (ii) substitution among goods and services based on estimated own- and cross-price elasticities of demand (Friedman and Levinsohn, 2002). In this particular case, the study found that substitution in consumption dampens the welfare losses from the Asian crisis by about 50 percent.

The Indonesian crisis has also provided a laboratory for understanding household responses on the income side of the picture. A study by Smith *et al.* (2002) offers a comprehensive analysis of changes in employment, wages and family incomes during the 1986–98 period, with a special focus on household responses to the crisis of 1997/98. They found that, while real wages were sharply reduced during the crisis – by as much as 60 percent in the case of formal sector employment in rural areas – combined family income in these rural areas fell by only about 37 percent during the crisis.

The dampening effect is attributed to the relatively stable returns to self-employment activities (primarily agriculture) and the increased allocation of family labour to self-employment. The study found that when the value of production for home use was included in the calculations, "full" family incomes (wages, plus self-employment income, plus production for home consumption) in rural areas fell by 21 percent, or about one-third of the decline in wages.

The urban households in Indonesia were not so fortunate. While urban wages fell by somewhat less than rural wages (55 percent), full family income in the urban areas fell by twice as much as in the rural areas (43 percent compared with 21 percent in rural areas) during the first year of the crisis. The relative increase in the price of food and farmers' ability to increase production in response to higher food prices were important factors in the rural households' ability to withstand the Indonesian crisis.

In fact, during this crisis, the agriculture sector demonstrated a remarkable capacity to absorb workers, with the farm labour force expanding by 20 percent (7.2 percentage points when measured relative to the entire workforce) during just one year. This flexibility in the face of external shocks suggests that considerable potential exists for such rural economies to adapt to, and benefit from, the higher world prices for agricultural products that are expected to follow multilateral trade liberalization.

Another way to assess the potential for developing countries to benefit from higher agricultural prices in the wake of trade liberalization is by estimating the agricultural commodity supply elasticity. Households gain from a price increase if they are net suppliers, but even if a household is not a net supplier prior to the reforms, given sufficient output response to the price hike, it might become a net supplier after the price increase. Thus, its chances of a welfare gain are considerably enhanced in the presence of large supply elasticities.

The evidence on agricultural supply response in developing countries suggests that the supply elasticities for individual crops are substantial, while those for the sector as a whole are quite small (Sadoulet and de Janvry, 1995). Infrastructure has a significant impact on supply response (Binswanger, 1989). The inability of the poorest households to increase production may be constrained by the lack of key productive assets (Deininger and Olinto, 2000). In summary, limited supply response can hinder the potential for such commodity price increases to pull households out of poverty in the absence of complementary policies aimed at improving access to credit and improved technology.

One study of the effects of agricultural trade reforms on poverty and inequality that takes into account both consumer demand and producer supply response to commodity price changes is that by Minot and Goletti (2000). In this study, rice production and consumption were subjected to a series of policy experiments, including (i) removing the rice export quota, (ii) changing the quota level, (iii) replacing the quota with a tax and (iv) removing restrictions on the internal movement of food. The aim was to understand how rice market liberalization in Viet Nam affects income and poverty in that country.

The distributional consequences of these policy scenarios were determined by way of the net rice sales position of different household classes, but these sales positions can change in response to how rice prices change. For instance, export liberalization raises prices within the country, particularly in the rice-exporting areas. The higher prices have a positive effect on rural incomes, and are generally favourable with regard to the number of people in poverty. Relaxing the restrictions on the internal movement of rice from south to north generates net benefits for the country, without increasing most measures of poverty.

Because rice production is relatively labour-intensive in Viet Nam, a rise in prices tends to increase demand for agricultural labour, and consequently the agricultural wage rate. Higher rice prices then lead to a greater decrease in poverty, particularly in households that derive a share of their income from agricultural labour. The counterfactual analysis in this work assumes that labour demand and wage rates remain constant because landlessness and the use of hired labour are considered not to be widespread in Viet Nam. However, as is clear from the next section, this is not necessarily the case in other countries.

Impact of trade reforms on factor markets

In the longer run, by stimulating the demand for unskilled labour in rural areas, higher agricultural prices tend to result in higher rural wages, thereby benefiting wage labour households in addition to self-employed farmers. Ravallion (1990) addresses this issue in a study of rural labour markets in Bangladesh that measures both the short- and long-run impacts of an increase in the price of rice on rural wages and poverty. A simple condition was used to determine whether such households gain from an increase in the price of rice. The condition required the elasticity of wages with respect to the price of rice to exceed the ratio of net food (rice) expenditures divided by net wage income.

On this basis, Ravallion concluded that the average landless poor household loses from an increase in the rice price in the short run, but gains in the long run (five years or more). This is because the increase in household income (dominated by unskilled wages) is large enough to exceed the increase in household expenditures, of which less than half is comprised of rice for the poorest households.

Two studies by Porto (2003a, 2003b) offer a natural generalization of Ravallion's work for the case of Argentina. Adopting a general equilibrium approach, a set of wage equations for unskilled, semi-skilled and skilled labour were estimated where the explanatory variables were international prices for all merchandise commodities (not just agricultural goods), educational attainment and individual household characteristics. The resulting wage–price elasticities were used to estimate the impact on wages of potential changes in domestic commodity prices arising from trade reforms.

These relationships were used to provide an *ex-post* analysis of the distributional consequences of the Southern Common Market (MERCOSUR) for households in Argentina (Porto, 2003b). The results, summarized in Figure 24, illustrate that MERCOSUR benefited the poorest households in Argentina substantially (6 percent of income), while the wealthiest households may well have lost (the dotted lines give the 95 percent confidence interval on these results). By removing policies that favoured the wealthy relatively more, MERCOSUR is estimated to have had a positive impact on the distribution of income in Argentina.

A separate paper by Porto (2003a) uses the same framework to conduct an *ex-ante* assessment of prospective reforms in domestic and foreign trade policy. In this case, he drew on outside estimates of the impact of foreign trade reforms on world prices. He concluded this work by noting that foreign reforms are more important than domestic reforms when it comes to potential poverty alleviation in Argentina.

Nicita's (2004) study of Mexican trade reforms referred to above uses the same approach as Porto to estimate how Mexican trade liberalization in the 1990s affected wages. Low-income households gained from lower-priced consumption goods, but these gains were largely offset by reductions in unskilled wages and agricultural profits. As a consequence, the poorest households gained much less than the wealthy ones. In fact, while all households appeared to have gained from the reforms, the wealthiest households gained three times as much as the poorest. These findings are summarized in Figure 25.

The preceding analyses are premised on the assumption that commodity price changes are eventually translated into factor market changes and that the subsequent changes in wages affect household welfare. However, in some cases, transaction costs may be high enough to preclude household participation in these markets (e.g. the

FIGURE 24
Impact of MERCOSUR on household real income in Argentina

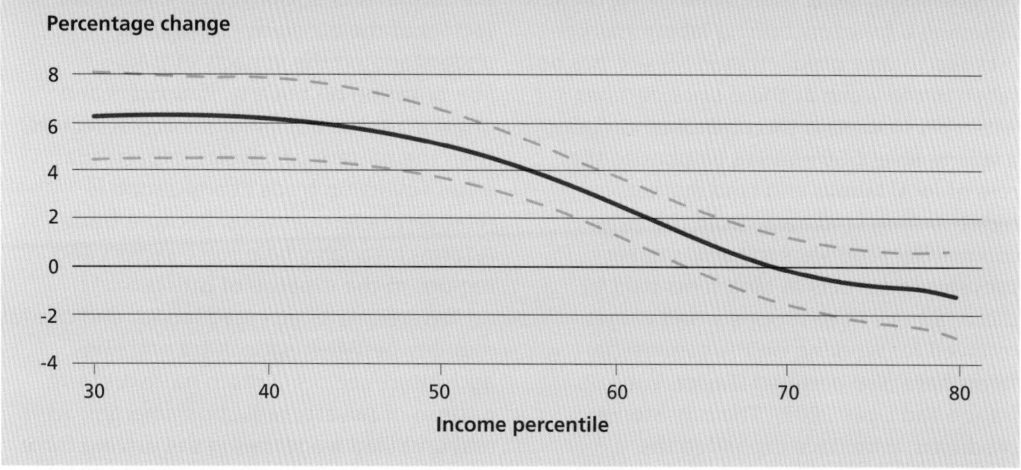

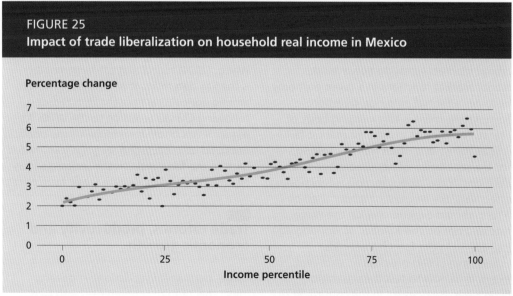

Source: Porto, 2003b.

FIGURE 25
Impact of trade liberalization on household real income in Mexico

Source: Nicita, 2004.

cost of travelling to the nearest job may be prohibitive). This factor can have effects that go well beyond the "missing market" itself.

A study of the role of market failure in peasant agriculture found that missing markets for labour and/or staple foods serve to dampen substantially the supply response of peasant households to changes in cash crop prices (de Janvry, Fafchamps and Sadoulet, 1991). This line of reasoning, coupled with the prevalence of subsistence producers in Mexico in the early 1990s led de Janvry, Sadoulet and Gordillo de Anda (1995) to conclude that the majority of the maize producers in the *ejido* or communal sector would be little affected by the declines in grain prices expected to occur under NAFTA. As a consequence, their estimates of the overall reduction in maize production were considerably smaller than those of the models assuming a fully functioning labour market.

In fact, maize production in Mexico has not fallen in the wake of these price declines. Attempts to explain this phenomenon using a village-level CGE analysis emphasize the role of local labour and land markets in redistributing land away from the large commercial producers towards smaller subsistence farmers as land rents paid by these farmers have dropped, and wages received for working on the commercial farms have also declined (Taylor, Yunez-Naude and Dyer, 2003). The subsistence producers, who have expanded the cultivated area, bolstered maize production in the wake of the price drops.

Given that the main endowment of the poor is their own labour, the market that merits greatest attention by those studying trade and poverty is clearly the labour market. Assessing how well the labour market functions in a given economy becomes a central empirical question. Fortunately, there is an emerging body of literature aimed at testing for market failure – or as the issue is often framed, testing for the separation of household and firm decisions. If the labour market is functioning effectively, the amount of labour used on a farm should depend only on the wage rate and not on the number of working-age individuals in the farm households.

Benjamin (1992) provides an excellent example of how to test the separation hypothesis. He does so, in the context of rice production in Indonesia, by incorporating demographic variables in the farm firm's labour demand equation and testing for the significance of the associated coefficient. Interestingly, he fails to reject the separation hypothesis, meaning that markets appear to be working.

However, the lack of wage labour income among many of the poorest rural households in some of the poorest countries suggests that this hypothesis might well be rejected in other cases. Hertel, Zhai and Wang (2004) note that nearly 40 percent of households in the poorest developing countries are completely specialized in farm income. These households are also disproportionately poor. Therefore, further examination of the separation hypothesis appears warranted.

The more general question of labour mobility – both across sectors and between the formal and informal (self-employed) sectors of the economy is crucial to understanding the impacts of trade liberalization on poverty. If workers and physical capital are immobile across sectors, then the pattern of poverty impacts that arises following trade liberalization is relatively heterogeneous, because trade reforms invariably help some sectors and regions at the expense of others.

However, with increased labour and capital mobility between agriculture and non-agriculture sectors, a much more uniform pattern of poverty reduction emerges, with real unskilled wages being the driving force behind these changes (Hertel *et al.*, 2003).

Recent econometric evidence from rural China suggests that the degree of off-farm labour mobility is quite low, particularly for households with low educational attainment (Sicular and Zhao, 2002). Hertel, Zhai and Wang, (2004) found that off-farm mobility is the key determinant of whether poverty among agricultural households is reduced following China's accession to the WTO. At higher levels of off-farm mobility, the boost in unskilled manufacturing wages is transmitted back to the farm and lifts the welfare of low-income households, despite lower farm prices.

Trade reforms, productivity and economic growth

Large, permanent reductions in poverty inevitably require economic growth (see Box 9). So the question naturally arises: to what extent will trade reforms stimulate such growth? There are numerous mechanisms through which this can work. Three possibilities are presented here: increased investment in physical or human capital, access to improved technology, and increased competition.

A recent study of Viet Nam's rice market reforms of the 1990s demonstrates that the resulting boost to agricultural prices and hence rural incomes enabled the rural poor to invest in human capital (Edmonds and Pavcnik 2002). The trade reforms that raised the price of rice, and hence rural incomes, substantially reduced the incidence of child labour, while simultaneously increasing the rate of school attendance. In fact, the rise in rice prices during the reform period of the 1990s explains fully half of the decline in child labour that occurred at this time. This is precisely the kind of effect that will result in long-run reductions in poverty.

Of course, this process can also work in reverse. The impacts of the Indonesian financial crisis on household spending resulted in substantial reductions in the amount allocated to education and health care in the wake of this external shock (Thomas et al., 1999). Moreover, the reductions were most pronounced among the poor. As Thomas and co-authors note, this reduction in human capital investment "suggest[s] that for these households the impact of the crisis is likely to be felt for many years to come".

Increased trade can also bring with it access to new technologies that can, in turn, have a significant impact on productivity. High trade barriers, both tariff and non-tariff in nature, often prevent access to some technologies or goods altogether, thereby impeding productivity growth (Romer, 1994). The case of maize production in Turkey provides a compelling example of the importance of imported technology (Gisselquist and Pray, 1997). Prior to 1982, Turkey restricted the importation of new varieties of agricultural commodities through a single-channel system, which gave the Ministry of Agriculture authority over seed production and trade. Between 1982 and 1984, these restrictions were relaxed, permitting foreign investment in this sector, the importation of new varieties and the elimination of price controls on seeds.

The impact on yields was dramatic. Comparing actual with predicted yields under previous technologies shows that these reforms contributed to a 50 percent increase in maize yields in Turkey. The increase in average returns to maize production was estimated at 25 percent of gross economic value.

There is also evidence that exporting can lead to enhanced productivity and that imports can effectively discipline domestic mark-ups in imperfectly competitive industries, thereby encouraging firms to move down their average total cost curve. In addition, many trade agreements have explicit components aimed at stimulating foreign direct investment (FDI), which can stimulate growth by adding to the host country's capital stock as well as bringing with it new technologies and managerial capacity.

For example, in a study of FDI, research and development, and spillover efficiency in Taiwan Province of China, Chuang and Lin (1999) used firm-level data to confirm the existence of beneficial spillovers from FDI. They found that a 1.0 percent increase in an industry's FDI ratio produces an increase of 1.4–1.88 percent in domestic firms' productivity.

Model-based evidence
Cline (2003) modelled the links among trade liberalization, productivity growth and poverty. Specifically, he combined econometrically estimated elasticities of growth with respect to trade, as well as the elasticity of growth with respect to poverty, with a CGE analysis of global trade liberalization. This permitted him to synthesize an estimate of the aggregate, long-run poverty reduction that might arise from such a policy change. Cline began with the global CGE model of Harrison, Rutherford and Tarr (1997), augmenting the static gains from trade (the focus of the studies cited above) with the "steady-state" quasi-dynamic gains that follow in the long run from increased investment.

To this, he added another pure productivity effect, which he inferred by multiplying the increase in trade for each region – as estimated by the CGE model – by a "central estimate" of the elasticity of output with respect to trade, distilled from a review of the now vast cross-country growth regression literature. With the estimate of long-run growth in per capita income resulting from

BOX 9
Why trade matters for reducing poverty and improving food security?[1]

Supachai Panitchpakdi, former Director-General, World Trade Organization

Technology and modern agriculture has transformed the nature of the quest for food security but in one respect there has been no significant change. Despite the impressive material progress that our civilization has made, hunger and starvation have sadly not been eradicated in all parts of the world.

Today there is the realisation that a sustainable domestic food supply cannot be ensured by each government acting individually. History has repeatedly shown that protectionism and isolation from world markets have never been the right answer. Food self-sufficiency is not equivalent to food security. The goal of self-sufficiency is illusory in today's world where a vast range of inputs constitute the full production equation. Nor is any country insulated from sudden adverse climatic effects which can dramatically reduce domestic agricultural output.

The WTO's contribution to efficient production is obvious and actually requires no elaboration. What is perhaps less obvious is the WTO's contribution to keeping the peace which is so vital to ensuring that supply channels remain open. Let us not forget that international trade conflicts have historically been a frequent cause of war, which jeopardizes directly people's access to food. The GATT/WTO system has, since 1948, provided a framework for the rule of law, peaceful negotiation and conflict resolution in international trade relations. Moreover, economic integration through trade provides a powerful incentive for political cooperation among nations. If I may quote from Montesquieu: "Peace is the natural effect of trade".

It is therefore no coincidence that the multilateral trading system is an essential pillar of the global political system. Stable trading relationships are vital not only for food security but also for global security. It is also no coincidence that more than two-thirds of WTO Members are developing countries. After all clear and strong rules are of particular value to smaller and less powerful nations.

The WTO also contributes in more specific ways to food security. Ensuring efficient production and distribution of food supplies is, however, only part of the food security equation. Hunger and malnutrition are almost always the result of poverty. While many other factors play their role, the vast majority of the hungry and malnourished suffer from inadequate income, not from inadequate food supplies. The poor often lack purchasing power even when food supplies are domestically relatively plentiful or are readily available through world markets. A real lack of food supplies due to war, civil strife or natural disaster is comparatively small.

Seen in this light, one of the most concrete ways which the WTO can contribute to improving food security is by providing the opportunity to raise income levels through economic growth. As is recognized in the Rome Declaration and Plan of Action – trade is a key element for food security – as it stimulates economic growth. It permits the efficient transfer of food supplies from surplus to deficit regions. It allows countries to become self-reliant rather than trying to become self-sufficient, regardless of cost.

Since 1948, tariffs in the industrialized world have been cut by more than 80 percent in eight successive rounds of negotiation, and a vast range of quantitative restrictions and bureaucratic controls have been removed. Since 1948, trade has grown faster than international output in all but eight years. Trade liberalization has also been an important stimulus for the expansion of knowledge, technology and capital.

The other major contribution that the WTO can make is, of course, in terms of the impact of trade policy on agricultural production. A common policy for governments seeking to enhance food security via self-sufficiency is to maintain high border protection and high internal prices to encourage domestic production. This, however, has adverse impacts on food security. High internal prices can act as a regressive tax. Poorer consumers tend to be hardest hit by high food prices. Reducing their purchasing power undermines their food security. Subsidies and other measures to induce production may also inadvertently benefit those members of the farming community, particularly rich farmers and landowners for example, who need it the least. It is clear that for these countries the pursuit of self-sufficiency will be an expensive, and arguably less than optimal, route to food security.

The distortion introduced by such policies also affects other countries. Its most direct effect is to curtail the agricultural exports of countries and regions where food can be produced at lower cost. This aspect is particularly important for developing countries. For many of these countries, including the poorest amongst them, how well they do economically depends on how well they do in agriculture. Of course, improvements in agricultural output and export performance depend on a wide range of factors outside the trade policy sphere. But it is widely accepted and understood that a further reduction of trade barriers and trade-distorting subsidies will help boost the economic performance of developing country agricultural producers.

The elimination of subsidies may, in the short-term, have terms-of-trade consequences for net food importing developing countries, as world prices have been kept artificially low for so many years. This is an important consideration and the special problems of net food importing developing countries deserve attention. The WTO provides some mechanisms to help. However, to address this problem in a definitive way we will need a broader response that involves the international development and financial agencies.

From a development perspective, the outcome of the Doha Round must be more ambitious than what was achieved in the Uruguay Round, and we are on track for an ambitious outcome. But I must stress that to reach this outcome we will need meaningful results across the board, but especially in agriculture. All WTO Members will have to show considerable flexibility to reach an outcome which is ambitious and at the same time achieves a balance between import sensitivities and export interests.

Let us not forget that food has always been an important element of trade, with markets integrated to a greater or lesser extent for thousands of years. But during the twentieth century, trade in basic foodstuffs was subjected to increasingly higher impediments. The Doha Round gives us the opportunity to reverse this trend. We have in the Doha Development Agenda an obligation we must live up to, not only as trade negotiators but also as representatives of governments that have committed themselves to meet the Millennium Development Goals and other vitally important international development initiatives. The longer the reforms are delayed, the longer the development gains are postponed. Food security is a complex matter. Enhancing food security requires initiatives and policy actions on many fronts, with trade being only one element among others. That being said the successful completion of the Doha Round from a food security perspective can only be viewed as positive. The path to food security is through integration and interdependence, not protection and autarchy.

---

[1] This box is extracted from the former WTO Director-General's speech to the High-Level Round Table on Agricultural Trade Reform and Food Security, held in Rome on 13 April 2005. The full text may be accessed at http://www.wto.org/english/news_e/spsp_e/spsp37_e.htm.

trade reform, Cline applied a country-specific "poverty elasticity" with respect to growth, based on an assumed log-normal income distribution for each region, to obtain his final estimate for poverty reduction.

The estimates are large, totalling nearly 650 million people – the bulk of these in Asia, where the absolute number of poor (based on a $2/day metric) is large and trade growth is relatively high following multilateral trade liberalization.

Cline's growth-based estimates of poverty reductions stemming from trade liberalization are considerably larger than those obtained by the World Bank Development Prospects Group (2003). These authors used a recursively dynamic CGE model to estimate the poverty reduction in 2015 arising from gradual global trade liberalization between 2005 and 2010. Like Cline, they used a poverty elasticity with respect to income (in this case uniformly assumed to be 2.0 – a high number based on existing evidence) to convert economic growth into poverty reductions. Unlike Cline, they tracked the accumulation of capital in response to increased investment, and the openness/productivity multiplier is also an explicit part of their model. They concluded that such trade reforms reduce the number of people living in poverty ($2/day) by 320 million – roughly half of Cline's estimate.

Cline's synthetic estimates – as well as those of the Development Prospects Group (2003) – highlight the potential for trade liberalization to have a substantial long-run impact on poverty. However, in order to arrive at this estimate, he had to follow a long and arduous path, crossing several research "minefields" in the process: "steady-state" CGE analysis, growth theory and cross-country regression analysis, in addition to the literature on income distribution and poverty.

It will be some time before these individual pieces are strong enough to support anything more than back-of-the-envelope estimates of potential long-run poverty impacts of trade reform. In the meantime, most of this literature will continue to emphasize the short- to medium-run income distributional impacts of trade reform on poverty resulting from comparative-static estimates of the ensuing commodity and factor price changes. To the extent that most policy-makers focus on this shorter time frame, and because short-run impacts are especially important for households facing extreme poverty, FAO believes this emphasis is justified.

Implications for policy research

Agricultural trade liberalization can have an important impact on poverty and inequality. Because the bulk of the world's poor live in rural areas where the dominant livelihood is farming, any trade reforms that boost agricultural prices and agricultural activity tend to reduce poverty. However, the specific impacts depend on a number of factors.

The extent of price transmission from the border to local markets can vary widely – even within a given country – as was seen in the case of Mexico. Poor infrastructure and high transaction costs serve to insulate rural consumers from world price rises, while penalizing exporters. Any policies aimed at reducing domestic marketing costs will enhance rural welfare and improve the chances of rural producers benefiting from trade reform.

The ability of households to adjust to the price changes flowing from trade reform also varies considerably across countries, localities and types of households. The more responsive households are to the price changes, the greater the chance that they will be able to gain from trade reform. If they can increase supplies of products whose price has risen, while reducing consumption of these same goods, then any initial losses will be lessened, and gains will be enhanced. Of course, their ability to increase supplies is likely to be greater if they have adequate access to capital assets and credit – something that is notably difficult for the poorest farmers.

In the medium run, labour markets play a strong role in determining the poverty impacts of trade reform. Net purchasers of agricultural commodities can gain from higher prices – provided these prices translate into higher wages and provided they have access to employment at these higher wages. In fact, the impact of trade reforms on unskilled wages is central to the poverty story. Hence the importance of domestic policy reforms aimed at improving the functioning of labour markets.

Long-run poverty reductions from trade reform hinge critically on economic growth. The impact of trade liberalization on economic growth is an area of intense research at present. Preliminary findings, based on the currently available empirical evidence on the trade–growth linkage suggest that this can be an important vehicle for reducing poverty.

Key findings

- Labour markets play a key role in determining the poverty impacts of trade liberalization. Net purchasers of agricultural commodities, who initially lose owing to higher prices, can ultimately gain if these prices translate into higher wages and more jobs.
- The dominant endowment of the poor is their labour, and the impact of trade reforms on unskilled wages is central to the poverty story, underscoring the importance of complementary domestic policy reforms aimed at improving the functioning of labour markets.
- Preliminary findings, based on the currently available empirical evidence suggest the trade–growth linkage can be an important vehicle for reducing poverty. As our knowledge about this linkage improves in the future, our ability to assess the long-run impact of trade reforms on poverty will be greatly enhanced.
- The potential for trade to contribute to poverty reduction and food security depends on effective investments in infrastructure, institutions, education and health.
- Removing taxes on agricultural exports and tariffs on agricultural inputs (machinery, fertilizers and pesticides) in developing countries would improve the terms of agricultural trade and help producers compete on international markets and in their domestic markets.
- Safety nets and food distribution schemes are essential to ensure that low-income consumers are not penalized by rises in the prices of food imports.
- For many developing countries, the positive food-security impacts of trade on non-agricultural incomes, especially jobs and wages, are the biggest promises of trade.

6. Trade and food security

FAO estimates the number of undernourished people in the developing countries at 815 million (Table 10). While two-thirds of the total numbers of undernourished are found in Asia, the highest prevalence is found in sub-Saharan Africa, where 33 percent of the population is undernourished.

These are disturbing numbers given the global community's commitment to food security concerns, its capacity to produce more than enough food for every human being and its power to use modern information systems to pinpoint exactly where food is needed and to mobilize rapid transport systems to move food quickly around the globe.

As discussed in previous chapters, increased openness to international trade is unlikely, on its own, to lead to major improvements in economic growth or poverty reduction, and the same is true for food insecurity. Complementary policies, including public investments in pro-poor growth strategies and safety nets, are crucial if trade liberalization is to support food security strategies. Expanding markets through trade can provide growth opportunities, encourage efficiency, and remove scale and scope constraints in the case of small, low-income economies with limited domestic markets.

This chapter examines the issue of food security and discusses how trade and trade liberalization can help promote food security.

Participation in international trade does allow countries access to larger markets for their products. At the same time, it provides access to larger and less-expensive food supplies than if they had to rely on domestic production alone. International trade can also be a powerful channel for technology transfer, without which the prospects for productivity increases are limited significantly.

It is this potential for international trade to promote economic growth, reduce inequality and improve food security that FAO wishes to promote. FAO recognizes that greater participation in international trade is a fundamental component of an overall mix of policies that foster economic growth and reduce poverty and food insecurity.

What is food security?

Food security exists when all people, at all times, have physical, social and economic access to sufficient, safe and nutritious food to meet their dietary needs and food preferences for an active and healthy life.

TABLE 10
Food and hunger indicators by region

	East Asia	Latin America and Caribbean	Near East and North Africa	South Asia	Sub-Saharan Africa	Developing countries
Per capita food consumption (kcal/person/day)						
1964–66	1 957	2 393	2 290	2 017	2 058	2 054
2000–02	2 874	2 848	2 975	2 397	2 247	2 659
Millions of people undernourished						
1990–92	275	59	25	291	166	817
2000–02	152	53	39	301	204	815
Millions of people in poverty ($1/day)						
1990	472	49	6	462	227	1 218
2001	271	50	7	431	313	1 089

Source: FAO, 2003a, 2004b and World Bank, 2005b.

FAO focuses on four dimensions of food security:
- *Availability* is determined by domestic production, import capacity, existence of food stocks and food aid.
- *Access* depends on levels of poverty, purchasing power of households and the existence of transport and market infrastructure and food distribution systems.
- Stability of supply and access may be affected by weather, price fluctuations, human-induced disasters and a variety of political and economic factors.
- *Safe and healthy food utilization* depends on care and feeding, food safety and quality, access to clean water, health and sanitation.

Vulnerability refers to the full range of factors that place people at risk of becoming food-insecure. The degree of vulnerability of individuals, households or groups of people is determined by their exposure to the risk factors and their ability to cope with or withstand stressful situations. Food insecurity is a complex phenomenon, attributable to a range of factors that vary in importance across regions, countries and social groups, as well as over time (Figure 26). These factors

FIGURE 26
Conceptual framework for food insecurity

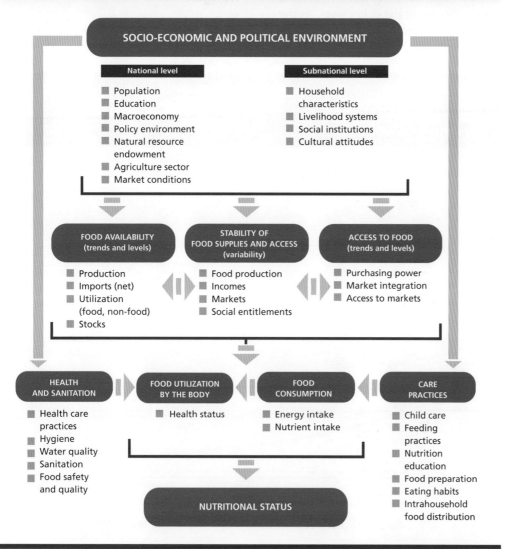

Source: FAO, 2000.

can be grouped in clusters representing the following four areas of potential vulnerability:
- the socio-economic and political environment;
- the performance of the food economy;
- care and feeding practices;
- health and sanitation.

To achieve success, strategies to eliminate food insecurity address these underlying causes by combining the efforts of those who work in diverse sectors such as agriculture, nutrition, health, education, social welfare, economics, public works and the environment. At the national level, this means that different ministries or departments need to combine their complementary skills and efforts to design and implement integrated cross-sectoral initiatives that must interact and be coordinated at the policy level. At the international level, a range of specialized agencies and development organizations must work together as partners in a common effort.

Trade influences these food security dimensions directly and indirectly through both general trade and agricultural trade in particular. For example, to the extent that increased participation in and integration into international trade fosters economic growth, increases employment opportunities and improves the income-earning capacity of the poor and food-insecure, it enhances access to food. In addition, openness to agricultural trade can promote food security by augmenting food supplies to meet consumption needs and reduce the variability of overall food supplies.

Correlations between trade and hunger

Increased integration of international markets has stirred widespread concerns

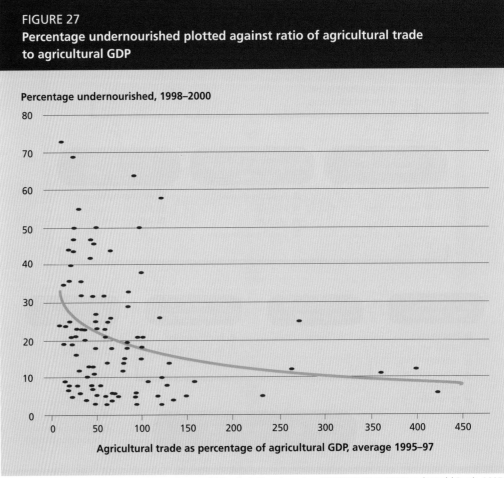

FIGURE 27
Percentage undernourished plotted against ratio of agricultural trade to agricultural GDP

Source: FAO calculations based on FAOSTAT; FAO, 2002 and World Bank, 2005c.

FIGURE 28
Percentage underweight plotted against ratio of agricultural trade to agricultural GDP

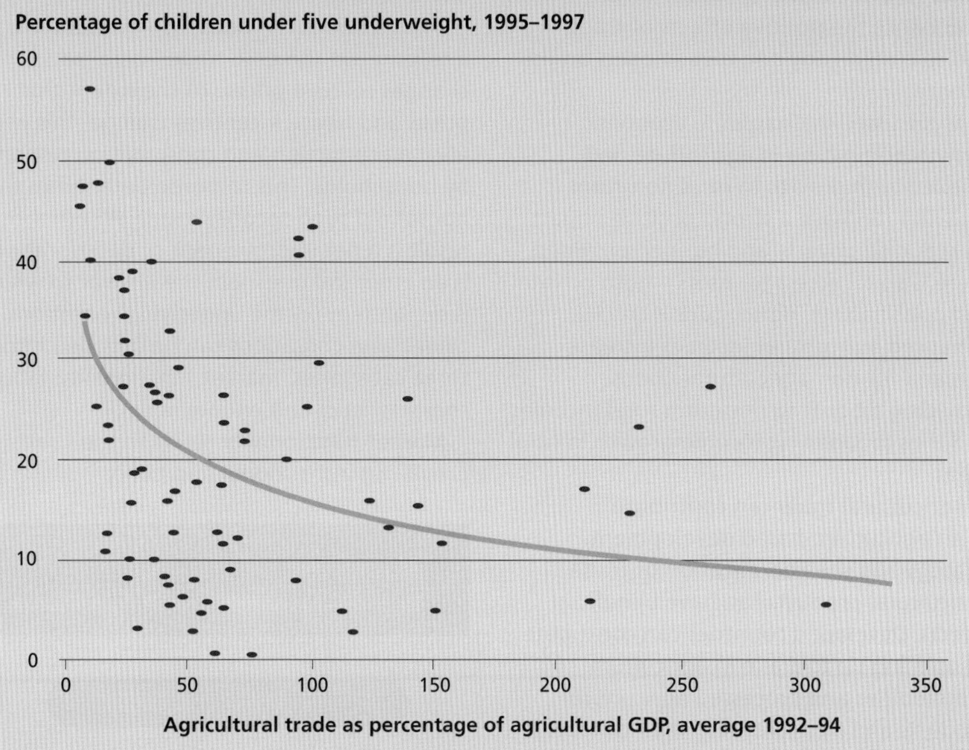

Percentage of children under five underweight, 1995–1997 (y-axis, 0 to 60)

Agricultural trade as percentage of agricultural GDP, average 1992–94 (x-axis, 0 to 350)

Source: FAO calculations based on FAOSTAT and World Bank, 2005c.

that openness to agricultural trade may jeopardize food security in developing countries. The concern is that exposure to international markets may increase the instability of food supplies and prices, disrupting markets and undermining incentives for local production. Is this fear supported by the evidence?

In Figures 27 and 28, agricultural trade (exports plus imports) as a share of agricultural GDP is plotted against two measures of hunger: (i) the proportion of the population that is undernourished and (ii) underweight prevalence in children under five years of age. Because changes in trade volumes would take time to have an impact on food security, the trade openness measure is lagged by a three-year period. If, indeed, agricultural trade were harmful to food security, then a high degree of agricultural trade openness would tend to be associated with a high proportion of undernourished people in the population. In neither of the two figures is this expectation borne out by the evidence. Without implying any causal relationship, the evidence does not suggest that engagement in agricultural trade is associated with high levels of undernourishment but, rather, the opposite.

However, another notable point that can be observed in the Figures is the wide degree of dispersion of the data. Each level of trade openness is associated with a wide range of hunger indicators. This suggests that the impact of agricultural trade and trade liberalization on food security is mediated by many other factors, such as markets, infrastructure, institutions and the complementary policy environment in which trade liberalization takes place. The importance of well-functioning markets, in particular, cannot be overemphasized.

The evidence does not point to a negative relationship between agricultural trade and food security; on the contrary, a higher degree of openness to trade is associated with lower levels of undernourishment. While this statement may hold in general, it is also true that some households lose in the process of trade liberalization, leaving their food security compromised. Hence, domestic policy reform must accompany trade reforms to enhance the positive effects of trade and to cushion any negative impacts on the hungry.

As discussed in Chapter 5, economic growth originating in agriculture and coupled with growth in rural non-farm activities can have a strong positive impact on reducing poverty and hunger, provided there is equal access to assets – both private and public assets. Agriculture is crucial. Enhanced agricultural incomes increase demand for non-agricultural goods, providing a boost to non-farm rural incomes and thus broadening income growth in rural areas.

Agricultural trade can contribute to an agriculture-based development strategy and the liberalization of trade in agricultural products can have beneficial effects. Developed countries can contribute by opening up to trade in agricultural commodities and processed agricultural products and by preventing their domestic farm-support programmes from placing subsidized commodities on world markets to the detriment of developing country producers. Developing countries, on the other hand, can ensure that their own trade regimes are as conducive as possible to stimulating growth in the agriculture sector.

The review of the empirical evidence on trade liberalization in Chapter 4 suggests that the largest gains to developing countries tend to come from their own trade-liberalizing measures and domestic reforms. In this respect, it is likewise important to note that the agriculture sector can also be strongly affected by protectionist policies directed at other sectors of the economy, as discussed in Chapter 3.

Trade liberalization and food security

While agricultural trade can conceivably play an important role in reducing poverty and food insecurity, the precise effects of trade liberalization on food security are nevertheless complex. At a conceptual level, the relationship between trade reforms and food security can be seen in a reform–response–result framework (Figure 29) (FAO, 2003b; Morrison, 2002; McCulloch, Winters and Cirera, 2001).

For a given set of underlying conditions, a reform – in this case trade liberalization – changes relative prices. To the extent that prices and hence incentives change, this will elicit a production and consumption response by households. This response determines the household food-security result. Whether this result is a net improvement or deterioration is an empirical question, the answer to which is dependent on the underlying conditions. Underlying conditions can be grouped into three categories: market functionality, labour characteristics and endowments.

Market functionality refers to the prevailing institutional and policy

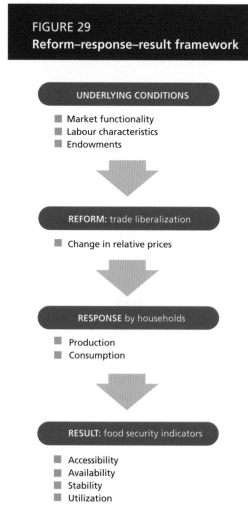

FIGURE 29
Reform–response–result framework

UNDERLYING CONDITIONS
- Market functionality
- Labour characteristics
- Endowments

REFORM: trade liberalization
- Change in relative prices

RESPONSE by households
- Production
- Consumption

RESULT: food security indicators
- Accessibility
- Availability
- Stability
- Utilization

> **BOX 10**
> **Cashew market liberalization in Mozambique**
>
> Mozambique liberalized its cashew sector in the early 1990s in response to recommendations from the World Bank. Opponents of the reform have argued that the policy did little to benefit poor cashew farmers while bankrupting factories in urban areas. Using a welfare-theoretic framework, McMillan, Rodrik and Welch (2002) analyse the available evidence and provide an account of the distributional and efficiency consequences of the reform. They estimate that the direct benefits from reducing restrictions on raw cashew exports amounted to $6.6 million annually, or about 0.14 percent of Mozambique's GDP. However, these benefits were largely offset by the costs of unemployment in the urban areas. The net gain to producers was probably no greater than $5.3 million, or $5.30 per year for the average cashew-growing household. It was estimated that the loss in real income to urban workers was around $6.1 million, which is roughly equivalent to the direct efficiency gain generated by liberalization. The apparent reason for the failure of the cashew reform was because it paid little attention to some key realities. First, traders and intermediaries rather than poor farmers captured most of the benefits. Second, because the world market for raw cashews is less competitive than that for processed cashews, Mozambique suffered a loss in its external terms of trade. Third, poor political management of the reform undercut the dynamic gains that could have resulted.
>
> The key to securing dynamic gains would have been a credible commitment to a new pricing regime – possibly complemented with compensatory programmes – that would have made the costly investments that were necessary worthwhile for farmers, entrepreneurs and workers. Liberalization could have reinvigorated the rural sector by reversing the collapse in cashew tree planting. In the urban sector, it could have heralded a restructuring of production by promoting more rational investment. However, farmers refused to plant trees, cashew processors refused to take their resources elsewhere and urban workers refused to look for other jobs.
>
> *Source:* McMillan, Rodrik and Welch, 2002.

environment, taking into account policy reforms other than the trade reform under way. It also incorporates physical and technical infrastructure such as transport and communication networks. Labour characteristics encompass human attributes, including education, health standards, asset ownership and the pre-existing level of food security. Endowments are a household's material attributes such as natural resources, climate, remoteness, land specificity and geographical proximity to borders. The case of cashew market liberalization in Mozambique (Box 10) illustrates the importance of underlying conditions to the success or failure of liberalization.

To the extent that prices actually change following trade reform (see the discussion of price transmission in Chapter 5), a farm household may have different possible supply (production) responses: intensifying or expanding existing production, diversifying or changing the input mix (e.g. using off-farm labour) or continuing production unchanged. Many of the poor and food-insecure are removed from formal market mechanisms; thus, in the absence of policies aimed directly at linking them to markets, they will most likely continue production unchanged. For such households trade reform alone cannot provide any beneficial effects.

The consumption (demand) response determines the food security result through the channels of access, availability and stability. A key question concerning the access channel is: what happens to a household's ability to buy food? This is affected by two indicators: income and the price of food. That is, are households able to produce or earn sufficient income to purchase the food they are unable to grow for themselves?

Availability is needed for households to be able to convert demand to consumption. Increased openness to trade will generally improve links to export markets, which, in turn, can be used to connect isolated rural poor to imported food produce.

Stability can also be affected by trade. A number of empirical studies (e.g. Anderson, 2000) have found that more open trade in agriculture improves price stability rather than aggravating it: if countries want the assurance of stable and predictable food supplies, they should seek more open trade, not more self-sufficiency. More open trade allows food to move from areas where it is in surplus to areas of deficit, and it enhances the capacity for deficit regions to feed themselves, both within and among countries (Runge et al., 2003).

The magnitude and direction of the relationship between trade liberalization and specific situations of food insecurity are empirical questions. While trade openness can contribute to improved food security, trade liberalization also has its costs. The resulting changes in relative prices and reallocation of productive resources can have a negative impact on some households' food security.

Tracing the impact of agricultural trade policy reform on household food security is not easy in practice. Many factors affect the way reforms feed through to the household level and there are likely to be both winners and losers. There are two dimensions to this. On the one hand, different typologies of households in different circumstances are likely to be affected in different ways. For some, the immediate effect of trade policy reform is likely to be beneficial; for others it may be negative.

Another aspect is the time dimension. While trade openness is expected to provide long-term benefits, many households face heavy costs, particularly in the short run. Indeed, policy reform imposes adjustments within countries as the incentive structure in the economy changes and productive factors are shifted towards those sectors in which the country has a comparative advantage.

Facilitating adjustment, while mitigating any adverse impacts on poor and food-insecure households, is a major policy challenge. Policies must be implemented that enhance the capacity of households to respond to the changed incentives. In this regard, access to well-functioning markets is essential – not least well-functioning labour markets. At the same time, compensation mechanisms and social safety nets are necessary to assist the negatively affected vulnerable households.

Trade policy reform offers opportunities to the poor and the food-insecure, but the adjustment process must be managed carefully with adequate protection of the vulnerable and food-insecure. The case studies examined below illustrate this point.

Case studies in macroeconomic and trade reforms

Countries that have liberalized agricultural trade during the last two decades under structural adjustment programmes and the AoA have experienced a range of food security outcomes. FAO recently carried out 15 country case studies[18] to examine the impacts of macroeconomic and trade policy reforms on food security (FAO, 2005a). A synthesis of the studies and the main findings are summarized in the following section and policy implications from the perspective of food security are drawn.

Structural characteristics of the sample countries

The structural characteristics of an economy, the existence and functioning of market institutions and the past policy context have an important bearing on the outcome of reforms and the appropriateness of alternative reform processes in country-specific contexts.

The countries in the FAO study sample are at different stages of development in the transformation from agrarian to industrialized economies. They vary from low-income agrarian economies (e.g. Malawi and the United Republic of Tanzania) with

---

[18] The case studies covered the following countries: in Asia: China and India; in Latin America: Chile, Guatemala, Guyana and Peru; in North Africa and the Near East: Morocco; and in sub-Saharan Africa: Cameroon, Ghana, Kenya, Malawi, Nigeria, Senegal, United Republic of Tanzania and Uganda. For a discussion on the methodology used, see also FAO (2003b, Chapter 11).

the majority of their populations engaged in agricultural activities, to predominantly middle-income countries with relatively low rural population densities (e.g. Chile and Peru). In the middle-income countries, structural transformations have already reduced the relative significance of agriculture to their national economies, to consumers and to rural incomes. India and, especially, China are special cases in this context because of their very large populations, their distribution between rural and urban areas and the absolute size of their industrial sectors.

Agriculture accounts for more than one-quarter of GDP in 9 of the 15 countries, exceeding 35 percent in 5 of them. Chile and Peru are the exceptions in that agriculture accounts for less than 10 percent of GDP in both cases.

In general, these structural characteristics imply differing impacts of liberalization on economy-wide effects and consumers' welfare, and of trade policy changes on agriculture and the rural population. For the low-income agrarian countries, the impact of agricultural growth, by virtue of the greater importance of the sector at their stage of development, is likely to be much more important for reducing poverty than in middle-income countries. For this latter category of countries, although the incidence of poverty is greater in rural areas, the absolute number of poor is small compared with that in urban areas.

Background and nature of the reforms

During the 1950s and 1960s, most of the governments of the countries in the sample intervened in their economies with the objective of accelerating the development process through rapid industrialization. The typical strategy pursued was one of import substitution, the counterpart of which in the agriculture sector being food self-sufficiency. In this regard, the countries in the sample pursued policies typical of those discussed in Chapter 3.

The fundamental challenges (and dilemmas) faced were (i) how to provide farmers with incentives to produce (i.e. remunerative and stable prices) while at the same time maintaining low prices of basic foods and agricultural supplies for the non-agricultural population (mainly urban dwellers/consumers) and (ii) how to correct for market failures (including missing markets) in the provision of basic services to the agriculture sector (e.g. regarding credit, essential inputs, technical and market information, and marketing and distribution infrastructure). Most of the governments intervened to influence both output and input prices and to provide basic services to the agriculture sector. In some cases, the intervention covered all agricultural products; in other cases it was confined to strategic products, either for domestic consumption or exports.

From a macro perspective, many of the countries in the sample had experienced periods of relatively rapid economic growth in the 1960s and 1970s before economic deterioration led to the need for policy reforms. The reforms that were implemented were often precipitated by a crisis in the economy signalled by both low growth and serious macroeconomic disequilibria – high inflation, fiscal deficits, current account deficits, and financial sectors in critical trouble associated in part with the foreign debt crisis of the early 1980s. Such constraints induced significant budget cuts generally and, more relevantly for agriculture, specific cuts in subsidized credit, marketing programmes and infrastructure.

In broad terms, the primary objective of the structural adjustment and liberalization programmes was to make domestic agriculture more market-oriented. The principal policy strategy adopted to achieve this objective was reform on several fronts: a reduction in average protection, deregulation, privatization and macroeconomic stability. The most important elements of policy reforms relating to agricultural trade were:

- the replacement of most quantitative restrictions on imports with tariffs;
- the reduction in both the level and dispersion of tariffs;
- the removal of export taxes, quotas and licences;
- a reduction in the importance of food self-sufficiency as a policy objective;
- a reduction in, or elimination of, state trading;
- the elimination of domestic price controls and the gradual removal of state procurement programmes.

TABLE 11
Average applied and bound MFN tariffs (percent)

	Applied MFN tariffs			Bound MFN tariffs		
	Year	Agricultural products	All products	Year	Agricultural products	All products
Africa						
Cameroon	1994	24	19	1998	80	–
	2002	24	18			
Ghana	1993	20	15	1995	97	92
	2000	20	15			
Kenya	1994	43	35	1996	97	96
	2001	23	19			
Malawi	1994	31	31	1996	111	76
	2001	16	13			
Morocco	1993	29	25	1997	66	43
	2003	52	33			
Nigeria	1988	37	34	1995	150	119
	2002	53	30			
Senegal	2001	15	12	1996	30	30
Tanzania, United Republic of	1993	28	20	1995	120	120
	2003	20	14			
Uganda	1994	25	17	1996	77	73
	2003	13	9			
Asia						
China	1992	46	43	2001	14	10
	2001	19	16			
India	1990	66	66	1996	115	49
	2001	42	32			
Latin America						
Chile	1992	11	11	1999	26	25
	2002	7	7			
Guatemala	1995	14	10	1999	51	38
	2002	11	7			
Guyana	1996	23	12	1998	93	58
	2003	23	12			
Peru	1993	18	18	1998	31	30
	2000	17	14			

Source: UN Comtrade database; World Bank/UNCTAD.

The sequencing and depth of reforms implemented varied across the countries and in some cases there were policy reversals. However, by the early 1990s, tariffs had been substantially reduced in most of the sample countries and were further reduced by 2001 (see Table 11). In the specific case of agriculture, average applied tariffs in 2001 were below 25 percent in all the sample countries except three – India, Morocco and Nigeria. Non-tariff barriers had largely been replaced by tariffs. However, a major issue that remained was the desire of governments to protect their agriculture sectors from world price fluctuations and to counteract export subsidies.

Consequences of the reforms for agriculture

Domestic price trends

The external economic environment of the past 20 years includes a downward trend in the real international market prices of many agricultural commodities, particularly maize, rice, coffee, cocoa, groundnuts and cotton. These declines have, however, been periodically reversed. Some products experienced more price fluctuations than others.

The movements in international prices are the result of many factors. For most tropical commodities, such as coffee, cocoa and tea, supply increases at the global level (due to increased productivity and the emergence of major new producers) have been the principal cause of the downward trend in international prices. However, for basic foodstuffs such as cereals, meat, dairy products and edible oils, which are typically import-competing in the sample countries, depressed international prices were mainly attributable to the high levels of domestic and export subsidies employed by developed countries. The international agricultural markets most distorted by high levels of support and protection included cereals (wheat, maize and rice), sugar, dairy products, meats and oilseeds.

In the absence of domestic policy measures aimed at maintaining agricultural prices, the downward trend in international prices translates to a downward trend in real farmgate prices. This can also apply to semi-tradeable products such as sorghum, millet, cassava and yams, whose prices tend to follow those of the major grains in the longer run.

In some cases, the reforms were associated with increases in most real domestic producer prices (e.g. Guyana, Nigeria, the United Republic of Tanzania and Uganda) in each period of reform. In other cases, there were periods of rising real producer prices and periods in which they fell (e.g. Cameroon and Kenya). In others, the reform periods were characterized by real price declines (e.g. Guatemala and Malawi).

The reasons for this heterogeneity in domestic price responses are complex, but the studies point to a number of key determinants. These can be broadly categorized as those that affect prices at the border and those that modify the price within the domestic economy, whether owing to direct price interventions or to institutional factors. The periods of rising real domestic prices were generally associated with real exchange-rate devaluations. The release of government controls over prices and marketing systems also led to gains in producer prices (especially for export crops) in some cases. On the other hand, import liberalization appears to have contributed to a decline in the real domestic prices of some commodities.

Two examples are illustrative. In Chile, a system of state controls over prices and marketing designed to keep food prices low for consumers was removed and the real exchange rate was devalued. In comparison with the pre-reform period (1964–73), these changes brought about significant increases in real domestic prices of all major farm products, including wheat, maize, beef and sugar, although all those prices exhibited a gradual downward trend during the 1990s. The reforms generally succeeded in improving the transmission of international price movements to domestic prices, with the exceptions of wheat and maize prices owing to the application of automatic adjustments in border protection brought about by the national safeguard (price band) mechanism (Figure 30).

In Ghana, the reforms were associated with declines in real domestic producer prices of import-substituting crops, including maize, rice and yams (Figure 31). The domestic producer price of cocoa, a major export crop, increased. The exchange rate liberalization appears to have benefited cocoa, but as a result of world price trends plus import liberalization, the opposite result occurred for import substitutes.

Production trends

There is some evidence that output has responded positively to real price increases and negatively to decreases; however, this was not always the case. The pattern of production response was found to be almost identical for export crops and for food crops. Of the 150 episodes for which data on both price and production changes are presented in the studies, in only 66 percent of cases

FIGURE 30
Evolution of real domestic prices and the real effective exchange rate in Chile

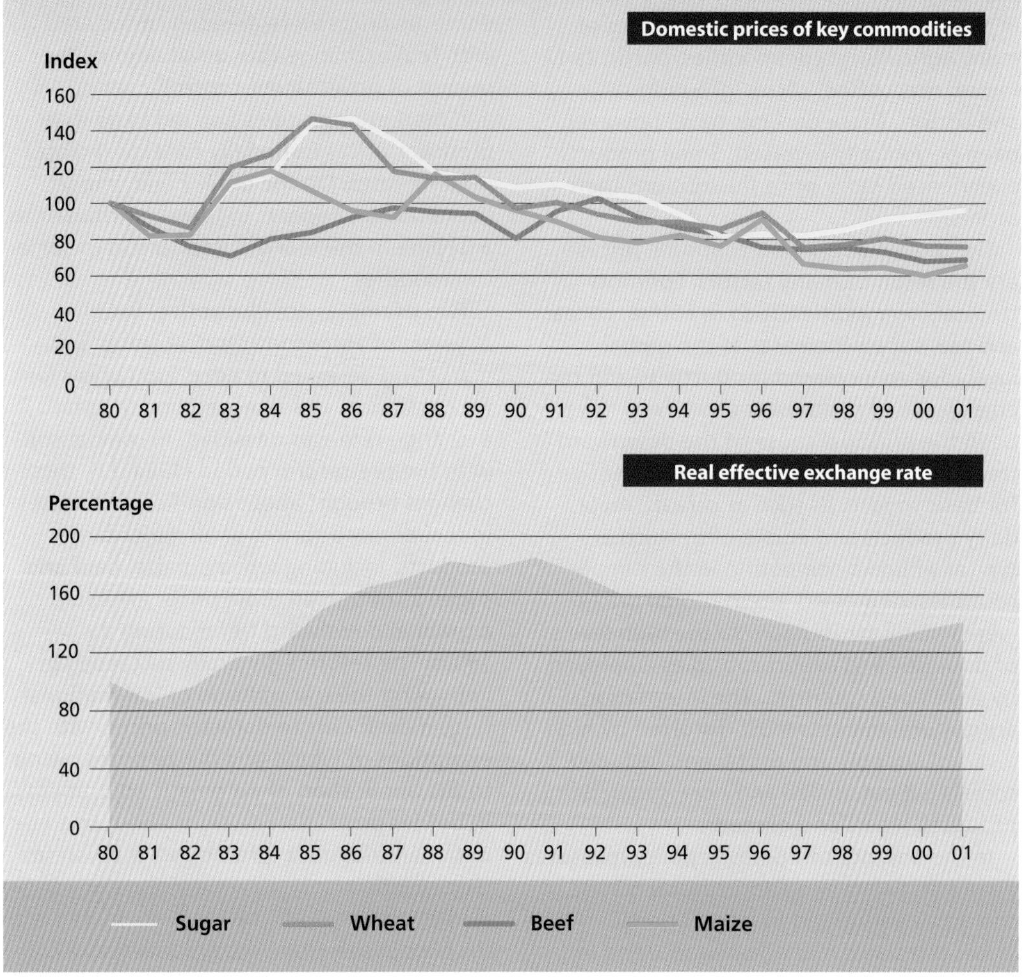

Domestic prices of key commodities (Index, 1980–2001)

Real effective exchange rate (Percentage, 1980–2001)

Sugar — Wheat — Beef — Maize

Source: FAO 2005.

is the response in the direction expected, with 34 percent of cases either reporting an increase in production when prices are falling or a decrease in production in the face of increasing prices. In Kenya and the United Republic of Tanzania, sectoral output fell in spite of real price increases. Malawi and Peru showed the opposite effect of increasing output across a range of products, in spite of declining prices.

Thus, overall, the picture is mixed regarding the apparent output response to price changes. This suggests that although producers respond to a combination of price incentives (determined both internationally and domestically), associated non-price constraints, or the alleviation of these, appear to play a critical role in determining whether a response occurs within the reform period (acknowledging that lags in response may explain in part these unexpected responses) and also the extent of the response.

As with the price changes, there are many reasons for the heterogeneity in production responses. Some have to do with changing world market conditions, as alluded to above. Where export opportunities increase as a consequence of the opening of a previously protected market, export expansion can occur despite falling international prices being more fully transmitted to domestic producers, as a result of concurrent reductions in local export restrictions.

Similarly, increases in domestic prices may

FIGURE 31
Evolution of real domestic prices and the real effective exchange rate in Ghana

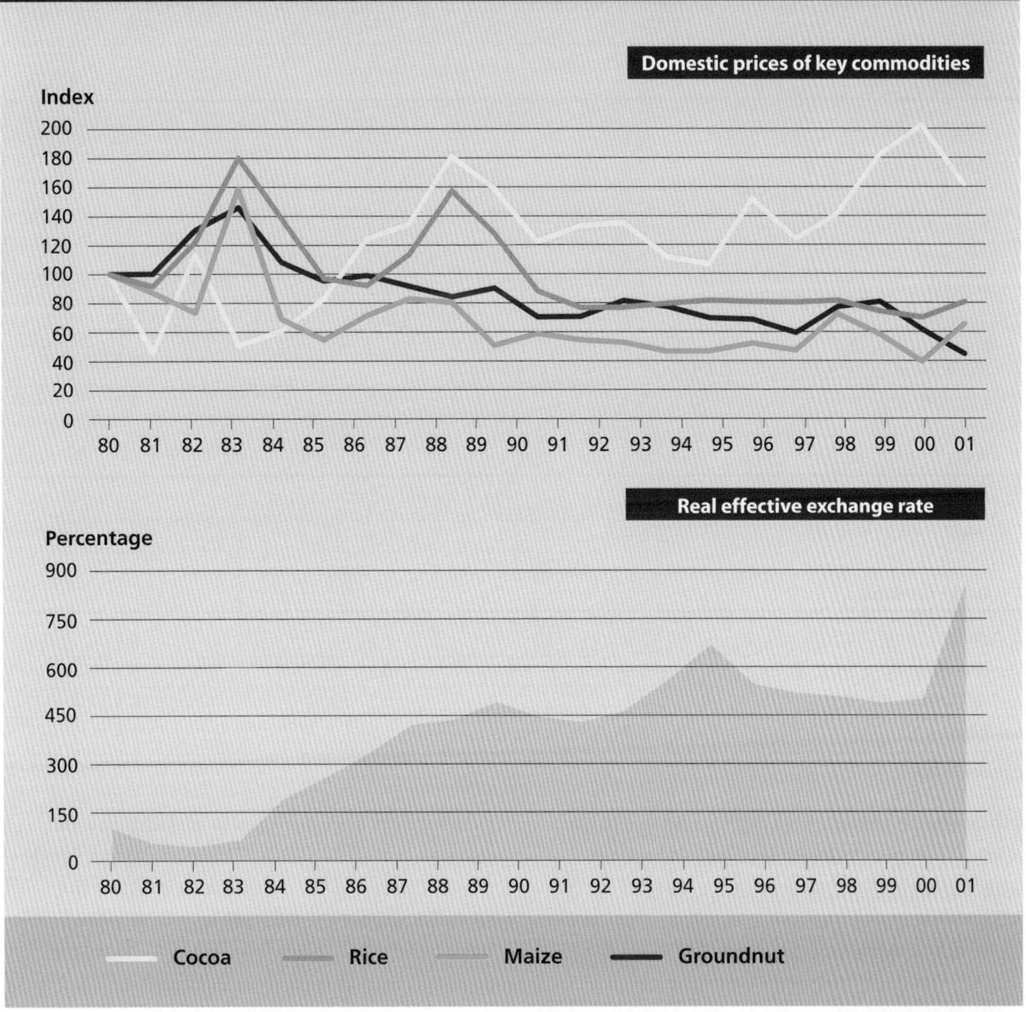

Domestic prices of key commodities

Real effective exchange rate

Cocoa — Rice — Maize — Groundnut

Source: FAO 2005a.

not reflect increased farmgate prices, as more powerful actors in the supply chain extract the increased rent associated with an increase in world prices. In other cases, domestic policy and institutional change can help explain the production response. For example, substantial rises in input prices may dampen the potential stimulus of increases in output prices; in other cases the withdrawal of support for rural credit may affect production negatively.

Trade trends
There were significant differences in the relative growth of agricultural exports and imports and, hence, in the direction of change of the ratio of food imports to agricultural exports among the sample countries (Table 12). After a period of strong growth in the first and/or second half of the 1980s, the growth rate of exports declined in the 1990s, falling sharply in the period 1992–97 for all countries except Chile and Senegal. In most cases, this pattern was associated with falling export prices rather than declines in volume.

In some cases not only did the total value of exports expand, but trade liberalization appears to have encouraged a diversification of export products. While the economic significance of traditional, tropical exports continued, there was a notable growth in non-traditional crops (e.g. fruits from Chile and Guatemala, cut flowers and string beans from Kenya or asparagus from Peru). In contrast, food imports generally rose –

TABLE 12
Ratio of total value of food imports to total value of agricultural exports

Country	Average ratio		
	1970–84	1985–94	1995–2002
Africa			
Cameroon	0.2	0.3	0.3
Ghana	0.2	0.4	0.6
Kenya	0.2	0.2	0.4
Malawi	0.1	0.2	0.2
Morocco	1.3	1.1	1.6
Nigeria	2.2	2.5	3.0
Senegal	1.2	2.1	3.7
Tanzania, United Republic of	0.2	0.3	0.5
Uganda	0.1	0.2	0.4
Asia			
China	0.8	0.5	0.7
India	0.6	0.4	0.5
Latin America			
Chile	2.8	0.2	0.3
Guatemala	0.1	0.2	0.3
Guyana	0.3	0.2	0.3
Peru	1.2	1.9	1.4

Source: FAO.

on average more quickly than agricultural exports in most cases. The net effect on the agricultural trade balance varied across countries.

In many African countries, food imports have increased not only because of reductions in border measures and exchange rate movements, but also because in the 1990s per capita food production fell in a number of countries (e.g. Kenya, Morocco, Senegal and the United Republic of Tanzania) (Table 13).

Effects of the reforms on food security
The implications of reforms for food security are difficult to gauge directly, and are best captured through a series of indicators that encompass both macro (national aggregate) and micro (household) characteristics. Such indicators can be categorized according to the three main facets of food security: availability, stability and accessibility.

Aggregate food supplies in the sample countries
There appears to have been a strong association between per capita food production and per capita availability of nutrients in the 1990s. The per capita food production index increased for 11 of the sample countries, including many significant improvements; however, Kenya, Morocco, Senegal and the United Republic of Tanzania experienced a worsening of this indicator that was particularly severe in the case of Morocco and the United Republic of Tanzania (see Table 13).

Among the countries whose food production index declined, only Kenya escaped simultaneous declines in nutrient availability. In other words, even if foreign exchange *per se* was not a limiting factor, other factors were evidently at work that prevented food imports from making up the production shortfall. A probable linkage in this regard operates via the effect of production on rural incomes, and the dependence of nutrition on income levels. Lack of sufficient income translates into lack of sufficient purchasing power to induce the marketing system to bring in needed quantities of imports.

The share of food imports in total imports rose in the period after 1995 in

TABLE 13
Changes in the proportion of the population undernourished, food production, rural poverty and economic growth

	Proportion of population undernourished			Change in proportion under-nourished	Real growth in per capita food production[1]	Incidence of rural poverty[2]		Real growth per capita[3]	
								GDP	Agricultural value added
	1979–81	1990–92	2000–02	1990/92–2000/02	1989/91–2001	Early 1990s	End 1990s	1990–2002	1990–2002
Africa									
Cameroon	22	33	25	–8	6	59.6	49.9	–1.2	2.0
Ghana	64	35	13	–22	48	63.0	49.0	1.9	0.7
Kenya	24	44	33	–11	–6	46.3	59.6	–0.7	–1.5
Malawi	26	49	33	–16	67	–	66.5	1.1	5.1
Morocco	10	6	7	1	–17	18.0	27.0	1.1	3.8
Nigeria	39	13	9	–4	18	48.0	76.0	0.3	0.9
Senegal	23	23	24	1	–3	–	–	1.0	–1.1
Tanzania, United Republic of	28	35	44	9	–22	41.0	39.0	1.1	0.8
Uganda	33	23	19	–4	1	59.4	39.0	3.6	1.0
Asia									
China	30	17	11	–6	74	32.9	3.2	8.2	2.9
India	38	25	21	–4	13	30.1	21.0	3.7	0.6
Latin America									
Chile	7	8	4	–4	25	39.5	23.8	4.2	1.6
Guatemala	18	16	24	8	3	–	–	1.2	0.1
Guyana	13	21	9	–12	84	45.0	40.0	3.5	3.8
Peru	28	40	13	–27	51	70.8	64.8	1.3	2.0

[1] Overall per capita food production growth between 1989/91 and 2001 in constant 1989/91 prices.
[2] Percentage of the population below the national poverty line. Starting and ending years differ for countries in the table, but are generally from 1990 to 2001, except for China where the beginning incidence of poverty is for 1978.
[3] Average annual percentage.
Source: FAO, 2005a; World Bank, 2005c; FAO, 2003b.

all countries in the sample except China, India and Peru (Table 14). A rising trend in this share reflects many factors, including population and economic growth, foreign exchange liberalization and relaxation of trade barriers. However, a rise in the ratio of food imports to total export earnings (goods and services) minus debt service suggests that food security at the national level (as reflected by the capacity to import) has become increasingly compromised. In the period following 1995, this ratio increased for eight of the countries in the sample, but declined for six others; it was particularly high (in excess of 20 percent) for Senegal, the United Republic of Tanzania and Uganda.

Household food security

Any impact of policy on the availability, accessibility and stability of food supplies at the national level is mediated by a range of institutional and regional parameters that affect what happens to individual households. Trends in household food security can be gauged from data on poverty and estimates of the proportion of undernourished.

Those countries that have experienced relatively strong rates of growth in real GDP per capita and/or in real agricultural value added per capita over the past decade tend to report positive outcomes with respect to the number of people below the national poverty line (see Table 13). However, it should be noted that the reductions in poverty have often varied across regions, and categories of farmers, in these countries.

For those countries experiencing relatively small increases in real GDP over the past

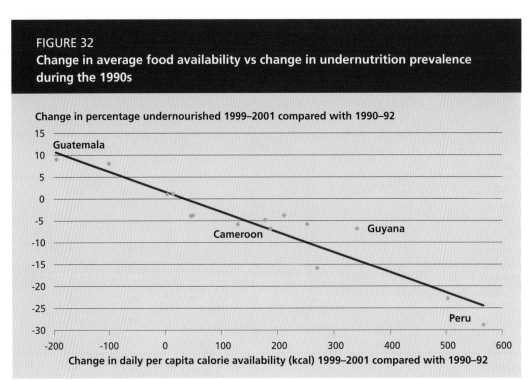

FIGURE 32
Change in average food availability vs change in undernutrition prevalence during the 1990s

Source: FAO, 2005a.

decade, the indicators for the poor are generally less encouraging. The case studies suggest therefore that the effects of the policy reforms on rural household incomes tend to depend significantly on the performance of the agriculture sector, including with respect to food production, and the overall response of the economy. This linkage reflects the relative importance of farm, off-farm and remittances in rural household income. In those countries in which the growth indicators, post-reform, were inadequate, there was a greater possibility that poverty would be exacerbated.

In addition to being closely associated with poverty levels, food security is reflected in data on undernourishment. Table 13 summarizes estimates of undernourishment and their trends. For most of the sample countries, the effects of reforms were felt between 1990 and 2001. In 2000–02, the United Republic of Tanzania had the highest prevalence of undernourishment, at 44 percent of the population; however, less than 10 percent of the population were undernourished in Guyana, Nigeria and Morocco and less than 5 percent in the case of Chile.

Over the period 1990–92 to 2000–02, FAO estimates of the proportion of the population undernourished declined in 11 of the 15 countries. The only countries for which undernourishment increased significantly were Guatemala and the United Republic of Tanzania. There is a strong correlation between changes in the prevalence of undernourishment during the 1990s and changes in average food availability, and in particular per capita food production (Figure 32 and Table 14).

Differentiated effects within countries
Within the agriculture sector of each country, reforms affected producers differently, depending upon cropping patterns. Producers of exports generally gained, as did wage-earners in production and processing in the export sector. In contrast, import-competing producers who lost some of their protection were generally adversely affected in the short run. However, their long-run welfare depended on their capacity to increase productivity and/or change cropping patterns. In many cases, farmers had little flexibility to adjust their production and output mix, and as a consequence the losses of this subset of farmers were probably long-term.

TABLE 14
Per capita availability of calories and protein, 1980/82–1999/2001

	Calories (kcal/day)			Protein (g/day)		
	1980–82	1990–92	1999–2001	1980–82	1990–92	1999–2001
Africa						
Cameroon	2 260	2 123	2 240	57	51	56
Ghana	1 661	2 094	2 621	38	46	54
Kenya	2 164	1 924	2 044	56	51	53
Malawi	2 269	1 886	2 164	66	51	54
Morocco	2 772	3 017	3 002	73	84	81
Nigeria	2 065	2 559	2 768	49	57	63
Senegal	2 343	2 283	2 275	67	67	63
Tanzania, United Republic of	2 186	2 078	1 970	54	51	48
Uganda	2 139	2 291	2 371	49	55	57
Asia						
China	2 400	2 708	2 974	56	66	85
India	2 067	2 368	2 492	51	57	59
Latin America						
Chile	2 646	2 612	2 851	71	73	78
Guatemala	2 332	2 352	2 160	59	60	55
Guyana	2 517	2 350	2 536	61	61	73
Peru	2 143	1 979	2 602	55	49	64

Source: FAO.

Producers of non-tradeable goods were generally less directly affected by trade reform, although they may have been harmed indirectly by consumers switching to lower-priced importables, or may have benefited indirectly from the higher price of exportables where such price rises occurred. Small farmers tend to be producers of non-tradeables, and their household members tend to be relatively more involved in rural non-farm labour. To the extent that increased employment opportunities became available in the rural non-farm economy, small farm households benefited from the reform process. Whether or not greater employment was caused directly by trade liberalization is, however, unclear.

There were also differences in real income effects on urban and rural consumers. It is well known that low-income households – urban and rural – spend a large proportion of their incomes on food. To the extent that trade liberalization lowers food prices, household income of the net-consuming poor may increase in real terms. Certainly, low-income consumers (small farmers are often net consumers too) benefited from trade liberalization as lower protection reduced the price of food relative to wage rates. This is clearly the case for those countries in the sample where farmers are a small proportion of the population (e.g. Chile) and/or most farmers are wage earners. However, where agriculture accounts for a large share of employment and farmers are self-employed, the loss of income for low-income farmers may outweigh any real income effects through the importables in their consumption baskets.

Summary of the consequences of reforms

Significant factors that influence the food security outcomes include the infrastructural and institutional context in which agriculture operates, the sequencing of reforms and the consistency of implementation of the reforms.

With the reduction of government controls over prices and marketing systems, macroeconomic reforms and the relaxation of trade barriers, particularly to exports, agricultural price incentives have improved in most, but not all cases in the countries studied. Changes in the real exchange rate, either as a result of exchange-rate policy or of inflation exerted a particularly strong influence on those incentives. When the exchange rate remained at an overvalued level, or appreciated following reforms, agricultural price incentives tended to deteriorate; the opposite was true when exchange rates depreciated.

Improvements in agricultural price incentives have generally led to increases in outputs (in two-thirds of the cases), but other factors dominated the incentive effect in the remaining countries. Withdrawal of available rural credit and a rise in input prices were among the reasons noted for a weak or negative supply response. In many cases also, when institutional reforms were undertaken to reduce government interventions in agricultural markets the private sector was not always fully equipped to replace government activities, resulting in an extended period of incomplete or inadequate services to producers.

Export agriculture has tended to benefit more from the reform packages than have import-competing crop sectors. While a dynamic export sector helps reduce poverty and food insecurity, producers in the import-competing sectors, especially small-scale producers whose assets are inadequate, may suffer losses, particularly when they are unable to switch to alternative productive or employment opportunities.

The rate of poverty reduction depends on aggregate economic performance (i.e. the creation of job opportunities, both on- and off-farm, at a rate that is faster than the rate of population growth). For largely agrarian-based economies, and where poverty is mostly rural, economic performance also depends to a great extent on the performance of the agriculture sector. Through this linkage, improvements in agricultural prices exert a positive effect on reducing poverty.

Food security, however, can be negatively affected by price increases if measures are not taken to ensure that smallholders and other poor rural households benefit in terms of real income improvement from the reform process.

Key findings

The underlying premise of the domestic and trade policy reforms undertaken by the countries studied was that greater market orientation would improve the sector's performance and ultimately promote poverty reduction and improve food security. The results from the reform experiences of the countries have been mixed. Nevertheless, a number of consistent themes emerged from the case studies.

- Reforms can be conducive to poverty reduction and improved food security if they are carefully designed and implemented within an explicit pro-poor strategy.
- The sequencing of reforms requires special and ongoing attention. Appropriate output incentives should be assured before (or at the same time as) input prices are raised, even at the cost of maintaining some well-targeted input subsidies during a transitional adjustment period.
- Finding mechanisms to encourage and assist the private sector to fill the gaps left by dismantling state agricultural marketing institutions is vital.
- Improving rural infrastructure is an important concomitant for successful policy reform in most countries, but it is particularly needed in low-income areas, along with support for productive investments by small farmers. Without such investments it is difficult for such farmers to respond to price incentives.
- Policies to encourage the development of rural non-farm employment are also important for the rural poor. These can include the development of microfinance, the simplification of regulatory regimes, infrastructure improvement, and special incentives for rural industrialization in poor areas.
- As complementary policies to facilitate adjustment of the kind mentioned above can take time to bear fruit, transitional compensatory measures, targeted at lower-income groups, may be needed. The absence of measures to protect the poor, and problems in targeting the most vulnerable groups, were noted in several of the case studies.

- Looking to the WTO negotiations on agriculture, the most sensitive domestic trade-policy debates centre on policy instruments to deal with import-competing sectors. This is particularly relevant in those cases where international markets are distorted because of high levels of support and export subsidies by wealthier countries that can afford them.
- For countries with a large proportion of low-income and resource-poor people living in rural areas and who depend on agriculture, reforms aimed at raising productivity and facilitating the transition out of agriculture and the creation of non-agricultural employment are essential for enhancing food security in the medium to long term. However, because such reforms may take some time to yield results, it is preferable that these reforms be set in motion before (or at least at the same time as) implementing measures such as removing subsidies on agricultural inputs and reducing tariffs on key crops grown by low-income households.

7. Making trade work for the poor: the twin track approach to hunger and poverty reduction

Dismantling trade restrictions is expected to provide long-term benefits for most countries at the national level. Over time, most people are able to take advantage of these emerging opportunities, but the adjustments and transition process takes time and the costs for many households may be heavy, particularly in the short run. Moreover, the impact of agricultural trade policy reform affects households very differently, depending on their location, ownership of assets, and how they earn their livelihoods.

Trade reforms must be accompanied by government policies to enhance the capacity of the poor to share the gains from trade and to compensate those who lose from the process, perhaps through social safety net programmes. The overall domestic policy environment is just as important as trade policies and must be conducive to private investment and private activity.

The importance of well-functioning markets is critical for reaping the benefits from trade liberalization and easing the adjustment. The price signals that reallocate resources are transmitted through markets. If key markets are missing or not functioning properly, these signals cannot be transmitted. A lack of good infrastructure – particularly roads, ports, telecommunications and marketing infrastructure – can constrain a country's ability to participate in and benefit from international trade.

Previous chapters reviewed evidence on the channels and pathways through which trade liberalization permeates down to households, ultimately affecting household food security. In summary, effective access to the trade-related opportunities is influenced in large part by the degree of price transmission, which, in turn, depends on the location of the poor, their access to infrastructure and transportation costs. The flexibility of households and enterprises in adjusting to trade-related opportunities is also highly dependent on factors such as their access to capital assets and to credit.

The work presented here suggests that effective labour markets are critical for determining how trade liberalization benefits can be spread beyond the immediate beneficiaries. Through its impact on general economic growth, liberalization can have significant, sustained long-run effects on levels of poverty and food security.

The main policy challenges for making trade liberalization work for the poor include:

- ensuring participation of all in the benefits of trade openness;
- facilitating the adjustment to a new set of relative prices;
- assisting people who are affected negatively.

Addressing these policy challenges requires a two-pronged approach. One line of action aims to ensure broad participation in the benefits and to facilitate the adjustment. A second line of action requires the establishment of direct assistance to provide food security to the groups who are affected negatively by the immediate impact of reforms. FAO's twin-track approach to poverty and hunger reduction provides an appropriate framework. Following an initial suggestion presented jointly by FAO, the International Fund for Agricultural Development (IFAD) and the World Food Programme (WFP) at the International Conference on Financing for Development in Monterrey, in March 2002, FAO first presented the twin track approach in more detail in its Anti-Hunger Programme (Box 11). In this framework, maximizing the contribution of trade, and agricultural trade in particular, to economic development is seen as but one, albeit indispensable, component of an overall development strategy for eliminating poverty and food insecurity. All measures proposed in this

BOX 11
Breaking the cycle of hunger and poverty: a twin-track strategy to reduce hunger and poverty

The extent to which the poor are able to take advantage of the opportunities provided by development depends critically on whether they are well nourished, in good health and literate. In particular, improvements in nutrition are a prerequisite for the poor to benefit fully from development opportunities. Hungry adults cannot efficiently perform physical and mental work, they fall sick more often and are likely to die young. Hunger perpetuates itself when undernourished mothers give birth to smaller babies who start life with a disadvantage. A vicious circle of hunger and poverty is created.

A twin-track approach is required for rapid success in reducing hunger and poverty:
(a) create opportunities for the hungry to improve their livelihoods by promoting development, particularly agricultural and rural development, through policy reform and investments;
(b) ensure direct and immediate action against hunger through programmes to enhance immediate access to food by the hungry, thereby increasing their productive potential.

FAO's "Anti-Hunger Programme", which was first released in draft form during the World Food Summit: *five years later*, held in Rome in June 2002, and in its final version during the FAO Conference in 2003,[1] sets out broad investment requirements and policy measures needed to implement the twin-track approach at the global level. It also proposes how the estimated extra public investment of $19 billion per year to enhance agricultural growth and productivity could be financed jointly by donors and recipient countries. Costs would be divided equally between the governments of countries where hunger is a problem and international donors. This would represent a doubling of concessional funding for agriculture from developed countries and an average increase of around 20 percent in total expenditures for agriculture by developed countries.

---
[1] FAO. 2003c. *Anti-Hunger Programme: a twin-track approach to hunger reduction: priorities for national and international action.* Rome (also available at ftp://ftp.fao.org/docrep/fao/006/j0563e/j0563e00.pdf).

---

strategy are fully compatible with the WTO Agreements.

Track one: creating opportunities

A domestic policy environment for poverty and hunger reduction

Stable and predictable macroeconomic policies are important for encouraging savings and investment, discouraging capital flight and promoting efficiency. Although many developing countries have moved towards macroeconomic stability, budget allocations for agriculture often remain painfully low. Substantial increases in budget allocations are particularly critical where hunger and poverty are prevalent and where the performance of agriculture, as the backbone of the economy, is well below potential.

Policy formulation and implementation should encourage participation by the poor and involve civil society organizations (CSOs) and the private sector. Administrative and fiscal decentralization makes it more likely that the poor will have a say in the decisions that affect them. It is important to enhance the functioning of markets through appropriate laws and regulations that ensure fair competition and safeguard market access by the poor.

Because agriculture is subject to a high degree of risk, it is also necessary to promote and improve instruments that address the need for risk management, especially with regard to the most vulnerable. This includes measures to ensure that markets for financial services allow rural populations to save, lend and borrow more efficiently.

Also, policies geared towards the rural economy must take account of the growing importance of non-farm rural activities, which offer the poor an escape route from poverty and constitute an integral part of their risk management and coping strategies. Policies and institutions are needed to develop rural infrastructure, build entrepreneurial capacity and ensure competitive and fair markets for small-scale rural enterprises.

Improve agricultural productivity in poor rural communities

Improving the performance of small farms in poor rural and peri-urban communities offers one of the best and most sustainable avenues for reducing poverty and hunger and providing a foundation for equitable economic growth. In addition to increasing and diversifying food supplies in local markets, it may also create a base for expanding and diversifying farm output into tradeable products and open employment opportunities.

Success in on-farm development depends on the creation of a policy environment conducive to agricultural growth, supported by research and extension institutions that are responsive to locally articulated needs. In many cases success also depends on developments beyond the farm boundary, such as improvements in road infrastructure or in the supply of irrigation water.

This process requires the emergence of self-reliant community institutions that can take the lead in ensuring the food security of all their members, plough gains back into new investments and develop linkages with other communities through the sharing of knowledge and experience. Associations of smallholders and rural community organizations, in coalition with CSOs, can play an important role in redressing some of the most serious disadvantages faced by their members and non-members. These include insufficient access to natural, financial and human capital; lack of access to appropriate technologies and income-earning opportunities; high transaction costs and insufficient access to markets; and lack of access to information, communications services and other public goods such as health and sanitation services.

Collective and coordinated action ensures greater responsiveness of the political process to the specific needs of communities and their members, prevents the abuse of pricing power for agricultural products and inputs by large buyers and sellers, allows producers to capture the considerable economies of scale existing in the procurement of inputs and marketing of outputs and facilitates the exchange of information and access to credit. The role of such partnerships and coalitions is particularly important in the face of government withdrawal from the provision of marketing services and credit.

Expand rural infrastructure and broaden market access

The rural areas of most developing countries still face inadequate levels of services and often a deteriorating stock of rural infrastructure. This infrastructural deficiency has resulted, *inter alia*, in reduced competitiveness of the agriculture of developing countries in domestic and international markets, and it has increased the costs of supplying growing urban markets from national farm production. Reversing the decline in the share of developing countries in world agricultural exports will require increased efforts by many developing countries to alleviate their domestic supply-side constraints. The highest priority must go to the upgrading and development of rural roads and to ensuring their maintenance, and to basic infrastructure to stimulate private-sector investment in food marketing, storage and processing.

Ensuring food safety and quality is an important factor in food security, as contaminated food is a major cause of illness and mortality. It is also important for broadening access to export markets. All developing countries are faced with an urgent need to invest in creating a stronger institutional capacity to ensure higher standards of food safety and quality and enable them to comply with international standards throughout the food chain. In an increasingly globalized market, it is also essential to take measures to prevent the spread of livestock and crop pests and diseases across national boundaries.

Post-production operations account for more than 55 percent of the economic value of the agriculture sector in developing countries and up to 80 percent in developed countries. However, relatively little public-sector and developmental support is targeted at this sector in developing countries. Action is urgently needed to develop food-handling, processing, distribution and marketing enterprises by promoting the emergence of small-scale farmers' input supply, processing and marketing cooperatives and associations.

A critical policy problem in the provision of infrastructure is how to address the relative neglect of poor rural communities. While involvement of the private sector in infrastructure construction and servicing may increase efficiency and respond better to overall needs, it may also mean that poor farming regions continue to be underserved. The public sector should maintain an active role in infrastructure provision that benefits the poor: decentralization and community participation in infrastructure investment planning, implementation, maintenance and financing should be encouraged to ensure demand-driven, sustainable service delivery and various forms of public–private partnerships should be considered.

Strengthen capacity for knowledge generation and dissemination

Rapid improvements in livelihoods and food security through on-farm investments depend on small-scale farmers having good access to relevant knowledge. Such access requires effective knowledge-generation and dissemination systems, aimed at strengthening links among farmers, agricultural educators, researchers, extension workers and communicators. Agricultural research and technology development are likely to be dominated by the private sector. There remain, however, many areas of basic research and, especially, extension where those who have not paid for the research cannot be prevented from enjoying its benefits. Private companies would be unwilling to conduct research in such areas, yet they may be vital for agricultural development and the sustainable management of natural resources. They include, in the case of research, most forms of pro-poor technology development and most approaches to farm development that do not depend on the increased use of purchased inputs – such as integrated pest management, measures to raise the organic matter content of soils or to improve fertilizer-use efficiency (e.g. through biological nitrogen fixation), or to conserve genetic resources. The responsibility for conducting research in these areas must, therefore, rest with the public sector.

National agricultural research and extension systems, many of which have deteriorated in their effectiveness, also need to increase their capacity to respond to the technology needs of small-scale farmers, particularly taking into account the acute labour shortage resulting from the HIV/AIDS epidemic in many African rural communities.

Policy action should aim at ensuring that the poor share the benefits of technological progress (in agriculture, information, energy and communications). This is particularly so for areas with poor agro-ecological potential, which are usually side-stepped by private commercial research. Public funding is required for the development and/or adaptation of technological options for those areas.

The emerging consensus is for a participatory approach to technology design and generation. Farmers' organizations, women's associations and groups and other CSOs can promote the necessary partnerships between farmers and scientists so that technological options are demand-driven and relevant. National policies should facilitate the establishment of functional linkages among research, extension education and communications.

Develop and conserve natural resources

With few exceptions, the scope for bringing additional natural resources into agricultural production (notably land and water resources) is limited. The only viable option is sustainable intensification, i.e. increasing the productivity of land, water and genetic resources in ways that do not compromise unacceptably the quality and future productive capacity of those resources. The policy environment must ensure that intensification is indeed sustainable and beneficial to the populations involved.

With regard to water, the key policy issue

is the growing competition between water requirements for agriculture and other water uses (domestic, industrial and ecosystem). As agriculture is by far the largest water user, the efficient use of water for agriculture should be the starting-point for expanding water availability for other uses. Policies affecting agricultural water use must provide incentives for efficiency gains and ensure that water scarcity is appropriately signalled to users. Transparent, stable and transferable rights to water use for individual users or groups of users are powerful instruments for promoting efficiency and distribution equity.

Concerning land for agricultural use, the most important policy issues are access and tenure (individual or community ownership, rental or longer-term user rights), improved land management practices and investments in soil fertility with a long-time horizon. Ensuring access to land contributes significantly to its sustainable use. In this context, strengthening women's rights to own and inherit land is particularly important. Policies should recognize the complexity of existing land tenure systems and of formal and informal arrangements regarding land-use rights. They should take into account the impact of increased mortality of the productive generation in rural areas as a result of HIV/AIDS and its potential impact on land-use patterns and inheritance arrangements.

Ensuring present and future access to a sufficient diversity of genetic resources for food and agriculture requires policy action at both the international and national levels.

For fisheries, the critical policy issue relates to limiting access to natural fish stocks where their capture, particularly marine, has reached or surpassed sustainable limits. Respecting limits on access to fish stocks requires that governments and fishing communities share authority and responsibility for making decisions about the use of fisheries resources.

In forestry, policies and institutions are needed to ensure full accounting of the value of the resource and benefits that accrue to the various members of society. These need to be incorporated into decision-making on utilization and conservation. Policies should encourage and promote the participation of key stakeholders in forest planning and management.

Track two: ensuring access

Beneficial trade liberalization hinges on getting food and income support to those who need it most. The principal objective of any food safety-net policy is the effective delivery of assistance to those in need.

It has been argued that trade barriers and price controls should remain on staple foods whose prices would rise following trade liberalization. However, this assumes that there would not be other foods that would be cheaper when opening to trade. Furthermore, there is no guarantee that those who are in most in need of food security support would actually benefit from import protection and price controls. Since the costs and benefits are opaque, such policy is likely to be captured by protectionist interests. Price controls and trade barriers are blunt – and often counterproductive – instruments in attempting to provide a food safety net for the poor.

A preferable policy for providing food safety nets for people who may lose from trade reform (in addition to those who are already hungry) is to provide direct food assistance. The advantage of direct assistance is that the costs (and benefits) of identifying the needy and distributing the food can be transparently assessed. The need to ensure direct access to food for the poor arises not only from humanitarian considerations and from the right to food, but also from the fact that it is a productive investment that can contribute greatly to fighting poverty.

Possible options for establishing food safety nets include:
- *Targeted direct feeding programmes.* These include school meals, the feeding of expectant and nursing mothers as well as children under five through primary health centres, soup kitchens and special canteens. Such schemes contribute to human resource development by encouraging children to attend school and improving the health and nutritional status of mothers and infants. They minimize nutrition-related illnesses and mortality among children, raise life expectancy and contribute to a fall in birth rates. Recent World

Health Organization estimates show that approximately 30 percent of children under five (approximately 200 million children) are more vulnerable to sickness and more likely to die early because of undernourishment.
- *Food-for-work programmes.* In many developing countries, a significant number of rural people are subsistence or below-subsistence farmers, producing only enough food to feed their families for part of the year. Food-for-work programmes provide support to such households while developing useful infrastructure such as small-scale irrigation, rural roads, buildings for rural health centres and schools.
- *Income-transfer programmes.* These can be in cash or in kind, including food stamps, subsidized rations and other targeted measures for poor households, and are also good means of increasing food-purchasing power and improving dietary intake.

Policies in this area should, *inter alia*, be derived from a human rights-based approach. A key policy prerequisite is the existence of information that identifies accurately who the hungry are and where they are located.

Programmes to provide direct assistance to the hungry can only succeed when national governments establish effective capacity for the delivery of such assistance. This requires a supportive national policy environment for the development of social safety nets, which can be provided in cooperation with CSOs.

Social safety-net policies specifically targeted at hunger reduction should give recognition to the special vulnerability of women and children to malnutrition at critical times in their lives and should support the creation and implementation of programmes such as mother–child feeding, related health and nutrition education and school feeding. A national commitment and relevant policies towards gender equality and women's rights is essential to enhancing access to food. At the household level, an improved status of women has been shown to be the most important single variable in reducing malnutrition.

Are we on the right path?

This report argues that trade and further trade reform can stimulate growth and have a powerful influence on reducing poverty and food insecurity, but that a strong domestic policy environment is necessary to ensure positive outcomes. This chapter has laid out an investment strategy aimed at securing the long-term benefits of trade reform in agriculture while protecting the weakest members of society from harm. So, are developing countries and the international community of aid agencies and donors on the right path?

The recent commitment of an increasing number of agencies and donors to assisting the developing countries in reaching the MDGs is encouraging. For example, as a part of their Africa aid strategy, G8 members preliminarily agreed in June 2005 to forgive $40 billion in debt owed by 18 of the world's poorest countries – 14 of them in Africa. Several donors have pledged to raise their development assistance to 0.7 percent of GDP. In May 2005, the EU detailed how it plans to reach this goal, announcing specific targets for the 15 older members of the EU as well as lower targets for the 10 newer members. Yet these commitments still have to be translated into concrete action targeting the poor and hungry.

The following section outlines briefly where we are in terms of investing in the long-term development of the agriculture sector in developing countries and promoting the sector's capacity for alleviating poverty and hunger.

Investment in agriculture lags where hunger is most prevalent

An overview of the data on private investment, public expenditures and external assistance to agriculture (EAA) in developing countries shows that the sector receives less investment and support in the very countries where hunger and poverty are widespread.

Most of the investment required to stimulate growth in the agriculture sector comes from private sources, mainly farmers themselves. A look at capital stock per agricultural worker in the primary agriculture of developing countries shows that it is extremely low and stagnant in countries

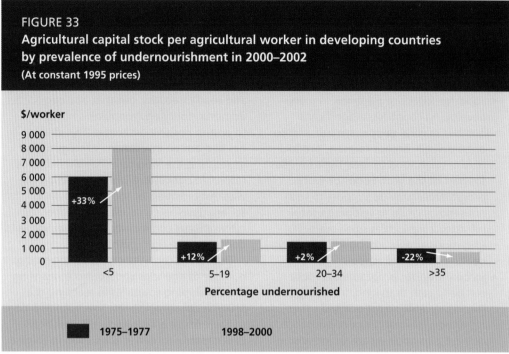

Source: FAO.

where prevalence of undernourishment is high, compared with those that have managed to reduce hunger.[19]

Moreover, this investment gap is growing. Countries with the lowest current levels of undernourishment (less than 5 percent of the population) have experienced strong growth in capital stock in agriculture since 1975. In all other categories, investment has increased little, if at all. And in the group of countries where more than one-third of the people are undernourished, the value of capital stock in primary agriculture has declined in real terms over the past quarter of a century (Figure 33).

Public investment fails to reflect the importance of agriculture

Public investment in infrastructure, agricultural research, education and extension is essential for stimulating private investment, agricultural production and resource conservation. The importance a government gives to agriculture compared with the importance of the sector in the economy can be illustrated by creating an "agricultural orientation index". This is found by dividing agriculture's share of public expenditure by its share of GDP. Figure 34 shows that countries with lower levels of undernourishment provide the strongest agricultural orientation in their public expenditure. In comparison, in countries with high levels of undernourishment, the index is very low. This clearly suggests the need for strengthening public expenditure on the agriculture sector in order to exploit fully its potential contribution to employment creation, poverty alleviation and reduction of food insecurity.

External assistance to agriculture does not target the neediest countries

External assistance is critical for very poor countries with limited ability to mobilize domestic private and public savings for investment. It is particularly critical for agriculture, which is largely bypassed by foreign private investors. Yet EAA declined at an alarming rate in real terms throughout the 1980s and stagnated in the 1990s. Despite pledges to increase aid, the most recent available data show no upward trend (Figure 35).

---

[19] Capital stock in agriculture refers to replacement value in monetary terms (at the end of the year) of tangible fixed assets produced or acquired (such as machinery, structures, livestock and land improvements) for repeated use in agricultural production processes.

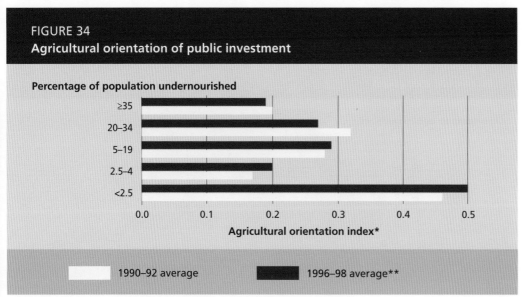

FIGURE 34
Agricultural orientation of public investment

* Share of agriculture in total public expenditures / Share of agriculture in GDP.
** Or most recent period for which data are available.

Source: FAO.

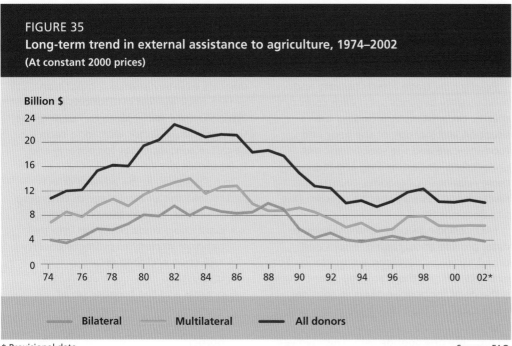

FIGURE 35
Long-term trend in external assistance to agriculture, 1974–2002
(At constant 2000 prices)

* Provisional data.

Source: FAO.

It also appears that EAA is not related to need. Data for 1998–2000 indicate that countries where less than 5 percent of the population were undernourished received three times as much assistance per agricultural worker as countries where more than 35 percent of the population were undernourished (Figure 36).

Summary

The common lessons, findings and insights, as well as the resolved and unresolved policy issues presented in this issue of The State of Food and Agriculture, reveal how trade–poverty linkages can be best used to

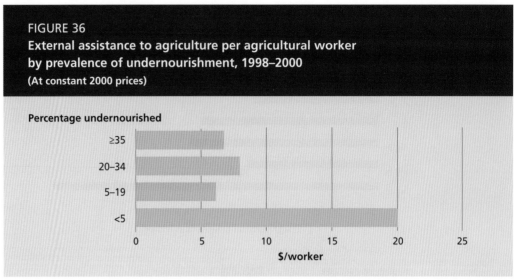

FIGURE 36
External assistance to agriculture per agricultural worker by prevalence of undernourishment, 1998–2000
(At constant 2000 prices)

Source: FAO.

enhance food security, address inequality and improve overall economic growth. The trade–poverty linkages are so complex, however, and national experiences with reform so different, that concluding with a set of unambiguous messages and stepwise policy guidance is an impossible task.

Exactly how trade affects poverty and food security depends upon each country's specific circumstances, including the situation and location of the poor and food-insecure and the specific reforms undertaken. Understanding and managing these relationships requires country-specific research and country-specific policies. One size does not fit all.

FAO's ongoing studies and analyses, to date, provide encouraging lessons and broad, strategic development guidance. For example, among the many important lessons is the need for policy-makers to consider more carefully than they have in the past how trade policies can be used positively to promote pro-poor growth. This involves actively implementing policies and making investments that complement trade reforms to enable the poor to take advantage of trade-related opportunities, while establishing safety nets to protect vulnerable members of society.

The analysis presented here concludes that multilateral trade liberalization offers opportunities for the poor and food-insecure by acting as a catalyst for change and by promoting conditions in which the food-insecure are able to raise their incomes and live longer, healthier and more productive lives. Trade liberalization also has adverse effects for some countries and households, especially in the short run as productive sectors and labour markets adjust. Opening national agricultural markets to international competition before basic market institutions and infrastructure are in place can undermine the agriculture sector with long-term negative consequences for poverty and food security.

To help minimize the adverse effects and to take better advantage of emerging opportunities, governments need to:
- clarify how trade policy fits into the national strategy to promote *poverty reduction and food security;*
- understand how the trade reform process and the broader set of economy-wide and sector-specific policies affect relative prices at the border;
- develop and use analytical tools to anticipate how trade polices may affect employment, local markets and distribution networks, and which economic and social sectors, which parts of the country, and which types of skills are affected.

To take advantage of the opportunities trade offers for pro-poor outcomes, consistent and sustained policy interventions are needed. Investments are required in rural infrastructure, human capital and other public goods, with priority being given to the expenditures and investments that are most essential to the poor and to the long-run viability of their livelihoods.

Safety nets are needed both to protect vulnerable groups from trade-related shocks and to allow the poor to take advantage of economic opportunities arising from trade. Of course, trade and trade reforms are not the only source of shocks faced by the poor and food-insecure. A host of other shocks – natural, human-induced and market-related – can spell disaster in the absence of effective safety nets.

Safety nets are not a substitute for addressing weak institutions, inadequate infrastructure and distorted factor markets, or for making essential investments in health, sanitation and education for rural people. As articulated by FAO's twin-track approach to hunger reduction, safety nets are an essential complement to these fundamental actions in preparation for more open markets.

Special contribution

Can trade work for the poor?
A view from civil society

The following is a contribution to *The State of Food and Agriculture 2005* by social organizations and movements, taken from their recent statements, evaluations and propositions regarding the liberalization of agricultural trade and its consequences. The International NGO/CSO Planning Committee for Food Sovereignty (IPC)[1] has facilitated this collective process, intended to focus on the food-insecure, the large majority of whom are vulnerable food producers.

This contribution, an autonomous and independent reflection on the issue of agricultural trade and poverty, in no way implies endorsement of the analysis published in the body of *The State of Food and Agriculture 2005*.

FOOD SOVEREIGNTY: A RIGHT FOR ALL ...
On 13 June 2002, during the final day of the FAO World Food Summit: *five years later*, summarizing the political analysis of the Forum for Food Sovereignty,[2] in the presence of the Heads of State and the Governmental Delegations, we stated:

> ... *Governments and international institutions have presided over globalization and liberalization, intensifying the structural causes of hunger and malnutrition. These have forced markets open to dumping of agricultural products, privatization of basic social and economic support institutions, the privatization and commodification of communal and public land, water, fishing grounds and forests ...*

> ... *We call for an end to the neoliberal economic polices being imposed by the World Bank, WTO, the IMF and Northern countries and other multilateral and regional free trade agreements ...*

> ... *We demand the removal of agriculture from the WTO ...*

The 2002 NGO/CSO Forum for Food Sovereignty, in its final resolution, outlined four major priority areas of action, one of which was the relationship between trade and food sovereignty. As stated in the NGO/CSO Forum action plan, "food sovereignty is a right of countries and peoples to define their own agricultural, pastoral, fishery and food policies which are ecologically, socially, economically and culturally appropriate. Food sovereignty promotes the Right to Food for the entire population, through small and medium-sized production, respecting: the cultures, diversity of peasants, pastoralists, fisherfolk, Indigenous Peoples and their innovation systems, their ways

---

[1] The IPC is a result of the non-governmental organization (NGO)/CSO and social movements process that developed the Forum for Food Sovereignty (Rome, 8–13 June 2002). The IPC is a facilitatory body that promotes and enables a debate with the United Nations agencies and international institutions based in Rome on agrifood-related policies. The IPC acts to enable discussions among NGOs, CSOs and social movements and will not substitute their direct relationships and negotiations. As stated by the FAO Director-General Jacques Diouf in the exchange of correspondence meant to identify the main lines of future relations between FAO and the IPC, "FAO recognizes the IPC as its principal global civil society interlocutor on the initiatives and themes emerging from the World Food Summit: *five years later* and the NGO/CSO Forum of June 2002."

[2] The Forum for Food Sovereignty, brought together in Rome, at Palazzo dei Congressi, from 8 to 13 June, more than 700 NGOs, CSOs and social movement representatives, including farmers, fisherfolk, pastoralists, indigenous people, environmentalists, women's organization, and trade unions, as the result of an international consultation and interaction process that lasted over two years.

and means of production, distribution and marketing and their management of rural areas and landscapes. Women play a fundamental role in ensuring food sovereignty." We now note that FAO's *State of Food Insecurity 2004* identified 80 percent of the most food-insecure people in the world as smallholder farmers, rural landless, pastoralists, fishers and forest-dependent men and women, thus acknowledging that current policies have failed in addressing the real causes of hunger.

The strategic approach on trade formulated in the 2002 NGO/CSO Forum plan of action states:

> *... globalization and liberalization have removed whatever limited support and protection exist for the majority of the world's farmers. It undermines local prices, undermines local producers to access their own markets. It forces producers to grow for distant markets at depressed and unstable prices. All subsidies to export-oriented agriculture have a dumping effect. [...] Trade should be a positive force for development, and should not contradict any human right. Food sovereignty requires fair agricultural trade, giving priority market access to local producers. Since its creation WTO did not apply food sovereignty principles. WTO is not transparent, democratic and accountable. Therefore, it does not have the legitimate position to deal with food and agriculture.*

SUMMARY OF THE "VIA CAMPESINA" POSITION ON TRADE

It is urgent to re-orient the debate on agriculture and initiate a policy of food sovereignty.[3]

Food and agriculture are fundamental to all peoples, in terms of both production and availability of sufficient quantities of safe and healthy food, and as foundations of healthy communities, cultures and environments. All of these are being undermined by the increasing emphasis on neo-liberal economic policies promoted by leading political and economic powers, such as the United States and the EU, and realized through global institutions, such as the WTO, the IMF and the World Bank. Instead of securing food for the peoples of the world, these institutions have presided over a system that has prioritized export-oriented production, increased global hunger and malnutrition, and alienated millions from productive assets and resources such as land, water, fish, seeds, technology and know-how. Fundamental change to this global regime is urgently required.

People's food sovereignty is a right

In order to guarantee the independence and food sovereignty of all of the world's peoples, it is essential that food is produced through diversified, community based production systems. Food sovereignty is the right of peoples to define their own food and agriculture; to protect and regulate domestic agricultural production and trade in order to achieve sustainable development objectives; to determine the extent to which they want to be self reliant; to restrict the dumping of products in their markets; and to provide local fisheries-based communities the priority in managing the use of and the rights to aquatic resources. Food sovereignty does not negate trade, but rather, it promotes the formulation of trade policies and practices that serve the rights of peoples to safe, healthy and ecologically sustainable production.

---

[3] Via Campesina, International farmers movement
(e-mail: viacampesina@viacampesina.org).

In this respect, market policies should be designed in order to:
- ensure adequate remunerative prices for all farmers and fishers;
- exercise the rights to protect domestic markets from imports at low prices;
- regulate production on the internal market in order to avoid the creation of surpluses;
- abolish all direct and indirect export supports; and
- phase out domestic production subsidies that promote unsustainable agriculture, inequitable land tenure patterns and destructive fishing practices; and support integrated agrarian reform programmes, including sustainable farming and fishing practices.

Trade rules must guarantee food sovereignty
Global trade must not be afforded primacy over local and national developmental, social, environmental and cultural goals. Priority should be given to affordable, safe, healthy and good-quality food, and to culturally appropriate subsistence production for domestic, sub-regional and regional markets. Current modes of trade liberalization, which allows market forces and powerful transnational corporations to determine what and how food is produced, and how food is traded and marketed, cannot fulfil these crucial goals. Trade in food can play a positive role, for example, in times of regional food insecurity, or in the case of products that can only be grown in certain parts of the world, or for the exchange of quality products. However, trade rules must respect the precautionary principle to policies at all levels, recognize democratic and participatory decision-making, and place peoples' food sovereignty before the imperatives of international trade.

The trade–poverty linkages
Export-oriented policies have resulted in market prices for commodities that are far lower than their real costs of production, perpetuating dumping. The adverse effects of these policies and practices are becoming clearer every day. They lead to the disappearance of small-scale, family farms and fishing communities in both the North and South; poverty has increased, especially in the rural areas; soils and water have been polluted and degraded; biological diversity has been lost; and natural habitats destroyed.

There is no "world market" of agricultural products
The so-called "world market" of agricultural products does not exist. What exists is, above all, an international trade of dumped surpluses of milk, cereals and meat. At present, international trade in agricultural products involves about 10 percent of total worldwide agricultural production, while the so called "world market price" is extremely unstable and has no relation to the costs of production.

Agricultural protection: of whom? How?
The larger parts of important agricultural and fisheries subsidies in rich countries are subsidies for corporate agri-industry, traders, retailers and a minority of the largest producers. This situation discredits agricultural subsidies in general which, in turn, negatively affects the possibility of maintaining much needed public financial support to peasant agriculture. Eliminating direct and indirect export subsidies is an important step but even more important is a policy to control supply. Supply management effectively eliminates surpluses. Effective supply management also

allows prices covering the cost of production and public financial support to peasant agriculture without generating surpluses that are dumped on other markets. Surplus-producing countries must limit their production and manage their supply in order to avoid excess production and subsequent dumping. These countries should orient their public assistance to the development of sustainable peasant production geared for the internal market. Importing countries should have the right to stop imports to protect domestic production and consumers; this should apply also to products with uncertain quality and safety such as genetically miodified organisms.

"Free" trade with "fair" competition is an illusion. Agricultural markets need strong state intervention.

By their very nature, agricultural markets cannot function in a socially just way without intervention by the state. Ending state intervention by eliminating agriculture policy instruments one by one would perpetuate the destructive restructuring of agriculture. This will displace millions and millions of men and women peasants, leaving them with no way to make a living. Regions and entire countries would be left with no capacity to produce food. Finally, only those who have money to purchase food will be able to eat. This scenario is catastrophic and includes an immense loss in terms of local varieties and food products, peasant knowledge and agricultural biodiversity.

AN AFRICAN PERSPECTIVE

Well-structured farmers and fisherfolks' organizations have emerged in Africa over the past decade. These organizations formulate visions and declarations which oppose liberalization of world agricultural trade and stress the ability of African agriculture to feed the region's citizens, as expressed in the *Afrique nourricière* campaign of the West African farmers' network, the ROPPA.[4]

The following contribution reflects the considerations which emerged from discussions on 2–3 February 2005, when representatives of peasants' and agricultural producers' organizations from Africa, Asia, Europe and the Americas met at the invitation of the ROPPA and the CNCR[5] to prepare their contribution to the International Forum of Dakar on the Global Agricultural Divide.

> ...*It is a common understanding that the neo-liberal policies and the dogmatic vision which have characterized development models of the past 20 years have ignored the basic missions of agriculture focused on food security, social equilibrium (job creation, limiting rural exodus, access to land, peasant or family-based agriculture, development of rural areas, etc.) and environment (soil quality, erosion, water). Moreover, they have also closed an eye to the imperfections of world markets. They have resulted in crises and an aggravation of the divide. The priority assigned to exportations has led to a collapse of world prices, to the deterioration of terms of exchange, without any benefits for consumers. It has also contributed to the ruin and the disappearance of a vast number of family farms in the South as well as in the North. A steadily growing number of peasants are landless, or lack access to means of production or financing.*
>
> *The solutions proposed by the Forum participants can be summarized as follows:*
>
> *1. **Reassign priority to the basic missions of agriculture.** This implies, in particular, respecting the following rights:*

---

[4] Réseau des Organisations Paysannes et de Producteurs de L'Afrique de l'Ouest (ROPPA) (e-mail: roppa@roppa-ao.org).
[5] National Rural Peoples' Council for Dialogue and Cooperation of Senegal (e-mail: cncr@cncr.org).

- *food sovereignty*
- *effective protection against importations*
- *access to means of production: water, land, seeds, financing ...*

2. **Stabilize world agriculture prices at a remunerative price** *for all farmers through mechanisms of regulation and supply management. Priority should be given to tropical product markets.*

3. **Introduce a moratorium on multilateral and bilateral agricultural negotiations (WTO and the United States Environmental Protection Agency [EPA])** *so long as they fail to respect the above principles...*

THE ARTISANAL FISHING SECTOR POSITION ON TRADE

Artisanal fishing, like other food-producing activities, is hard hit by adjustment policies, privatisation and liberalization of marine resources.

The following is the WFF[6] contribution to the *The State of Food and Agriculture 2005*.

Trade by itself does not contribute to people's development

...Small-scale fisheries have often been forgotten when international issues regarding food security and food sovereignty, and local and international trade are discussed. Their role as a source of economic income for coastal states at national and international scale is also neglected. This invisibility of small-scale fisheries made it possible, more than in other sectors, for the advocates of free trade to put pressure on governments to start privatizing national fishery resources, sometimes transferring them to transnational fisheries companies. As much as 99 percent of the catches from small-scale fisheries have a value as a commercial commodity or for direct human consumption.

The role of small-scale fisheries in international trade
Ensuring food sovereignty

Fish plays a fundamental role in feeding the world population since it supplies an important proportion of the protein consumption of hundreds of millions of people. Almost 16 percent of the world's average total consumption of animal protein comes from fish.

Preventing WTO rules from being applied to fisheries

WTO is a superpower which enforces international trade rules that facilitate the loss of sovereignty of states and nations. As a result, fisheries becomes an exchange currency comparable to other national economic sectors.

There is a need for international regulations on fish trade emanating from agencies of the UN system, like FAO and, more precisely, the United Nations Convention on the Law of the Sea (UNCLOS). An international agreement on subsidies and differentiated custom tariffs should be reached within multilateral bodies rather than in WTO.

Improving international agreements and treaties

Fish resources are the patrimony of nations and states are mandated to ensure their sustainable management. Thus states are not allowed to transfer the property of resources to third states and much less to international consortiums.

---

[6] World Forum of Fish Harvesters and Fishworkers (e-mail: forum@ccpfh-ccpp.org).

A GEOGRAPHICAL PERSPECTIVE: THE WESCANA REGION
The IPC brings together views expressed by different constituencies and regions. The following is a contribution which represents the view on trade of the WESCANA[7] region IPC representatives.

> ...Within the WESCANA region, the national governments have agreed to participate in all the regional and international trade agreements, and the various countries are at different stages of negotiation, signature or ratification. The majority of the countries are not exercising the negotiation process fully and they are practically accepting the liberalization terms without any major reservations, if any. Despite claiming that there are forms of grace periods for joining the free trade areas, there are no serious measures taken to ensure the ability of the local markets to withstand the impacts and cope with the competition caused by new barriers such as quality issues and indirect perverse subsidies.
>
> On the other hand, there is no consultation process with the affected groups accompanied by an awareness scheme to prepare them for the post-access phase. There is a very limited knowledge of the content and impacts of the various regional international and regional trade agreements on livelihoods and food sovereignty.
>
> It is well known that the countries of this region do not have the means to compete with more advanced countries and provide their farmers with a similar support.

---
[7] WESCANA Region – West and Central Asia and North Africa countries.

Part II

WORLD AND REGIONAL REVIEW
Facts and figures

Part II

1. TRENDS IN UNDERNOURISHMENT

- FAO estimates the number of undernourished people in the world in 2000–02 at 852 million: 815 million in the developing countries, 28 million in the countries in transition and 9 million in the developed market economies (Figure 37). More than half of the total number of undernourished, 61 percent, are found in Asia and the Pacific, followed by sub-Saharan Africa, which accounts for 24 percent of the total.

- The proportion of the population that is undernourished varies among the different developing country regions. The highest prevalence of undernourishment is found in sub-Saharan Africa, where FAO estimates that 33 percent of the population is undernourished. This is well above the 16 percent undernourished estimated for Asia and the Pacific and 10 percent estimated for both Latin America and the Caribbean and the Near East and North Africa.

- At the global level, the long-term trends for many food security indicators have been positive. The world total calorie supply per person has grown by 19 percent since the mid-1960s to reach 2 804 kcal/person/day in 2002, with the developing country average expanding by more than 30 percent. As consumption has increased, diets have shifted towards more meat, milk, eggs, vegetables and oils and away from basic cereals.

- The number of undernourished people has declined over the long run, although progress has slowed in recent years (Figure 38). The prevalence of undernourishment in developing countries fell from 37 percent of the total population in 1969–71 to 17 percent in 2000–02 (Figure 39). However, due to population growth, the decline in absolute numbers of undernourished people has been slower than that of the prevalence of

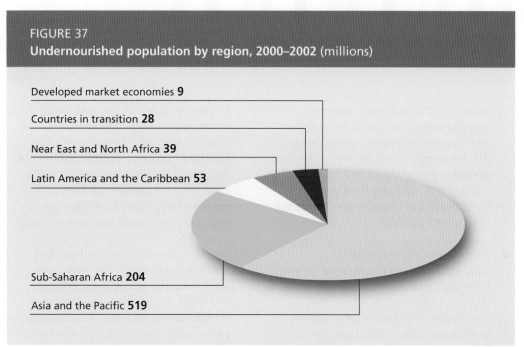

FIGURE 37
Undernourished population by region, 2000–2002 (millions)

Developed market economies **9**
Countries in transition **28**
Near East and North Africa **39**
Latin America and the Caribbean **53**
Sub-Saharan Africa **204**
Asia and the Pacific **519**

Source: FAO.

FIGURE 38
Trend in number of undernourished in developing countries, by region

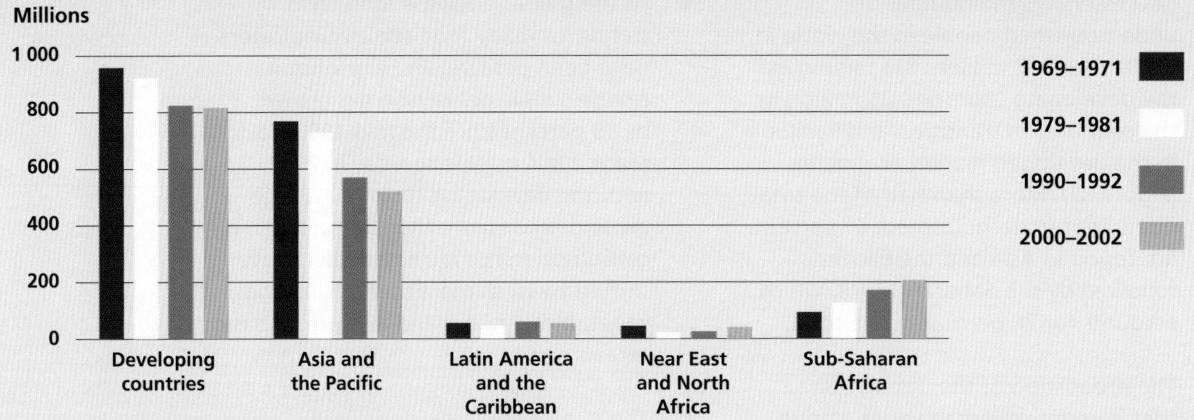

Source: FAO.

FIGURE 39
Trend in percentage of undernourished in developing countries, by region

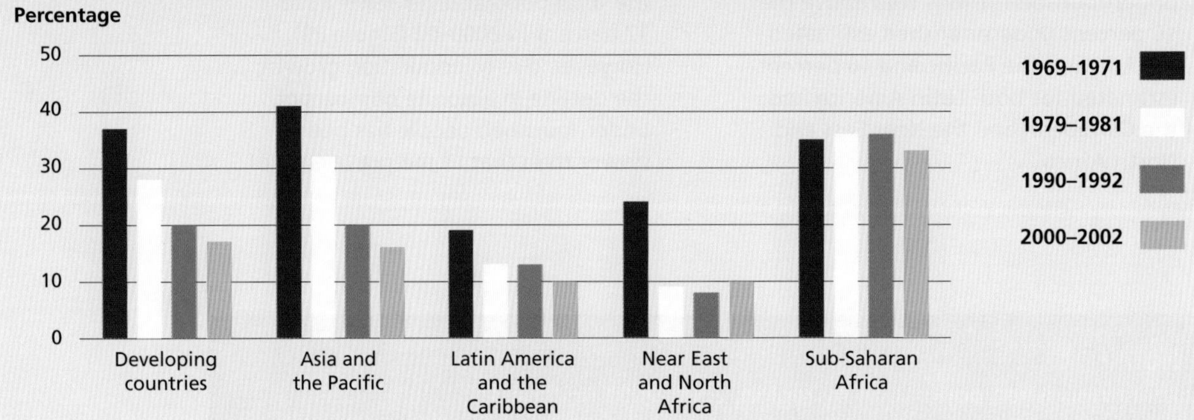

Source: FAO.

undernourishment. Past progress in aggregate food consumption numbers and undernourishment indicators for the developing countries was influenced decisively by the significant gains made in the most populous countries, including Brazil, China, India, Indonesia, Nigeria and Pakistan.

- Most of the improvement in undernourishment figures over the past three decades has been concentrated in Asia and the Pacific, where the prevalence of undernourishment has been reduced by almost two-thirds. In sub-Saharan Africa, the extremely limited reduction in the prevalence of undernourishment has been more than counterbalanced by population growth, resulting in a large increase in absolute numbers of undernourished people. Latin America and the Caribbean experienced a significant decrease in both prevalence and absolute numbers in the 1970s, but has made little progress since then. In the Near East and North Africa, the prevalence of undernourishment fell significantly in the 1970s, but by 2000–02 it stood at slightly above the level of two decades earlier, after having actually increased over the 1990s.

2. FOOD EMERGENCIES AND FOOD AID

- As of February 2005, the number of countries facing serious food shortages throughout the world stood at 36, of which 23 were in Africa, 7 in Asia and the Near East, 5 in Latin America and 1 in Europe. The causes are varied but civil strife and adverse weather predominate. A recent outbreak of desert locusts in western Africa and the tsunami disaster in South Asia have had serious though localized food security consequences. In many of these countries, the HIV/AIDS pandemic is a major contributing factor.

- Civil strife and the existence of internally displaced people or refugees were responsible for more than half of the reported food emergencies in Africa as of February 2005. The proportion of food emergencies that can be considered human-induced has increased over time. Indeed, conflict and economic failures were cited as the main cause of more than 35 percent of food emergencies between 1992 and 2004, compared with around 15 percent in the period 1986–91. In many cases, natural disasters are compounded by human-induced disasters, leading to prolonged and complex emergencies.

- The recurrence and persistence of emergencies often intensify the severity of their impact. In the period 1986–2004, 33 countries experienced food emergencies during more than half of the years. Many conflict-induced complex emergencies, in particular, persist to the extent that they develop into long-term crises. No fewer than eight countries suffered emergencies during 15 or more years within this period; in all instances, war or civil strife was a major contributory factor.

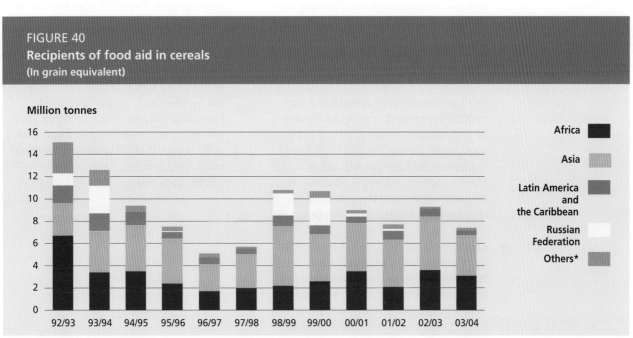

FIGURE 40
Recipients of food aid in cereals
(In grain equivalent)

Source: WFP.

* Including countries in transition
Note: Years refer to the 12-month period July/June. Countries of the Near East in Asia are classified with Asia. Countries of the Near East in North Africa are classified with Africa.

TABLE 15
Per capita shipments of food aid in cereals
(In grain equivalent)

	92/93	93/94	94/95	95/96	96/97	97/98	98/99	99/00	00/01	01/02	02/03	03/04
						(kg per capita)						
Africa	10.1	5.0	5.0	3.4	2.3	2.7	3.0	3.4	4.4	2.6	4.3	3.7
Asia	0.9	1.1	1.2	1.2	0.7	0.9	1.5	1.2	1.2	1.1	1.3	0.9
Latin America and the Caribbean	3.4	3.4	2.4	1.2	1.2	1.0	1.9	1.5	1.1	1.4	1.4	0.7
Russian Federation	7.6	16.7	0.1	0.5	0.1	0.3	13.6	16.8	2.1	1.1	0.0	0.2
Other	3.1	1.5	0.7	0.4	0.4	0.2	0.4	0.6	0.3	0.4	0.2	0.3

Note: Years refer to the 12-month period July/June. Countries of the Near East in Asia are classified with Asia; countries of the Near East in North Africa are classified with Africa.
Source: WFP.

- In contrast, many countries that possess relatively stable economies and governments but are consistently affected by unfavourable weather have implemented crisis prevention and mitigation programmes and established effective channels for relief and rehabilitation efforts. For these countries, a natural disaster need not result in a prolonged humanitarian crisis.

- Food aid in cereals fell to 7.4 million tonnes in 2003/04 (June to July), 1.8 million tonnes (or 20 percent) below the level for 2002/03. The sharpest decrease occurred in Asia, with shipments falling from 4.8 to 3.6 million tonnes – a decline of 25 percent. Other types of food aid increased slightly in 2004, but remain much smaller than cereal food aid (Figures 40 and 41). The top five recipients of cereal food aid in 2003/04, ranked in terms of volume of shipments, were Iraq, Ethiopia, the Democratic People's Republic of Korea, Zimbabwe and Bangladesh. All of these countries, except Zimbabwe, also counted among the top five food aid recipients during the previous year.

- Cereal food aid has been characterized by relatively large annual fluctuations. It has tended to decline relative to the levels of the late 1980s and early 1990s, but remains significantly higher than during the mid-1990s. Also in per capita terms, shipments have declined substantially relative to the early 1990s (Table 15). Disregarding exceptionally large shipments to the Russian Federation in certain years, Africa remains the largest recipient in per capita terms, albeit at levels well below those of a decade ago.

- The FAO Principles of Surplus Disposal and Consultative Obligations, originally agreed in 1957 and enshrined in the WTO Agreement on Agriculture disciplines on export subsidies in 1995, are intended to limit the potential of food aid to disrupt normal trade flows. Food aid may be further disciplined in the ongoing Doha Round of trade negotiations. The WTO Members have agreed to eliminate by a fixed date food aid that is not in compliance with operationally effective disciplines. The role of international organizations with regard to the provision of food aid by Members, including related humanitarian and developmental issues, are being addressed in the negotiations, as is the question of providing food aid exclusively and fully in grant form (WTO, 2004b: para. 18).

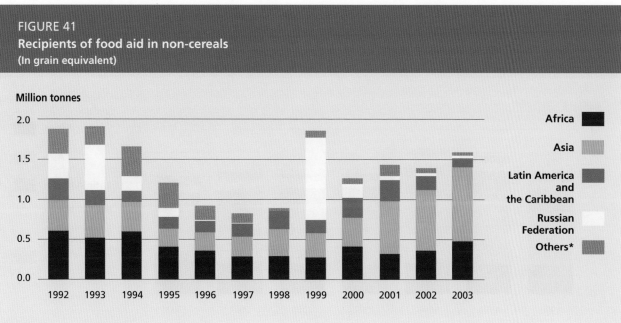

FIGURE 41
Recipients of food aid in non-cereals
(In grain equivalent)

* Including countries in transition
Note: Countries of the Near East in Asia are classified with Asia. Countries of the Near East in North Africa are classified with Africa.

Source: WFP.

3. EXTERNAL ASSISTANCE TO AGRICULTURE

- Measured in constant 2000 prices, preliminary data for 2002 indicate that total external assistance to agriculture was virtually unchanged from the previous two years (Figure 42). The global picture, however, masks shifts among the regions. Latin America and the Caribbean and the transition countries experienced variability in recent years while assistance to Asia continues to decline. External assistance to sub-Saharan Africa is estimated to expand for the third consecutive year, rising from $2.8 billion in 2001 to $3.4 billion in 2002.

- When measured in terms of amount per agricultural worker, external assistance to agriculture has more than halved since the peak level in 1982 (Figure 43). Among the developing country regions, sub-Saharan Africa seems to have recovered from the declining trend of the past two decades, with $17 of external assistance per agricultural worker in 2002. A possible slight increase in the 2002 values for the Near East and North Africa and Latin America and the Caribbean cannot be confirmed until final data become available. The amount of assistance per agricultural worker in Asia and the Pacific remains below that of other regions.

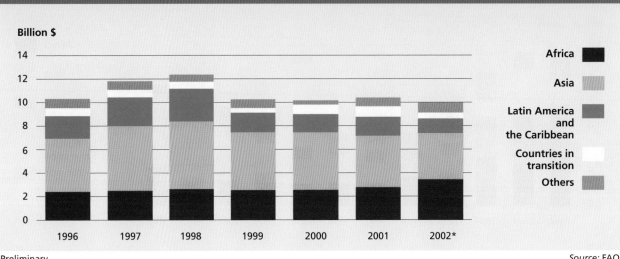

FIGURE 42
Commitments of external assistance to agriculture, by main recipient regions
(At constant 2000 prices)

*Preliminary

Source: FAO.

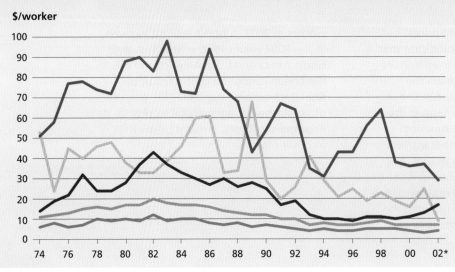

FIGURE 43
External assistance to agriculture per agricultural worker
(At constant 2000 prices)

*Preliminary

Source: FAO.

4. CROP AND LIVESTOCK PRODUCTION

- Global crop and livestock production increased over the past two years at rates above the averages of the previous four decades. The higher global agricultural output growth in 2003 and 2004 is the result of an upsurge in production in developing and developed countries, respectively. For the developing countries as a group, output growth peaked in 2003, but decreased in the following year to values below the averages of the previous decades. The developed country group recorded a significant output growth of almost 5 percent in 2004 after several years of contracting agricultural production. This rise is the result of a strong recovery in the transition countries and an increasing output growth in the developed market economies (Figure 44).

- In all developing country regions, output growth was lower in 2004 than in 2003. In Asia and the Pacific, agricultural performance improved in 2003, expanding by 4.5 percent after the lower 2 percent recorded in 2002. The rate of agricultural output growth in the region nevertheless declined in 2004 to 2.5 percent. Latin America and the Caribbean experienced systematically increasing rates of production growth over the period 2000–03, which slowed down to 2.4 percent in 2004. In the Near East and North Africa, agricultural performance continues to be characterized by pronounced fluctuations caused by variable climatic conditions in many countries in the region. After increasing by almost 7 percent in 2002, output growth will be virtually stagnant for 2004. Sub-Saharan Africa has likewise recorded variable growth in agricultural output over recent years compared with relatively favourable rates during most of the 1990s. Data for 2004 indicate an increase of only 0.5 percent in the region's overall agricultural production.

- Long-term trends in per capita food production provide an indication of the contribution of the sector to food supplies (Figure 45). Global per capita food production has increased steadily over the past 30 years, with an average annual growth rate reaching 1.2 percent during the past decade. Both the developing and developed country groups shared in this expansion, with per capita production growing at higher rates in the developing countries vis-à-vis the developed countries.

FIGURE 44
Changes in crop and livestock production

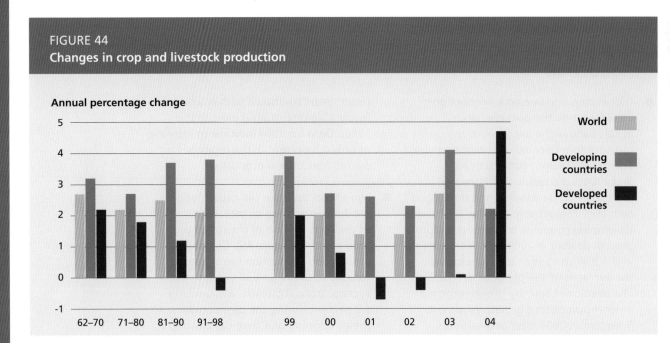
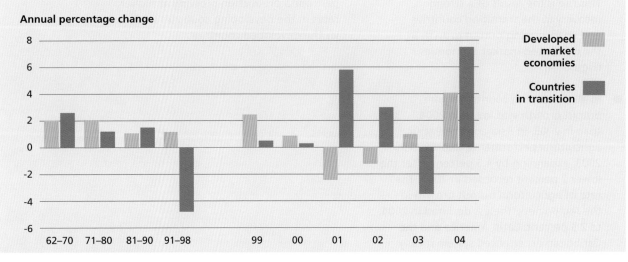
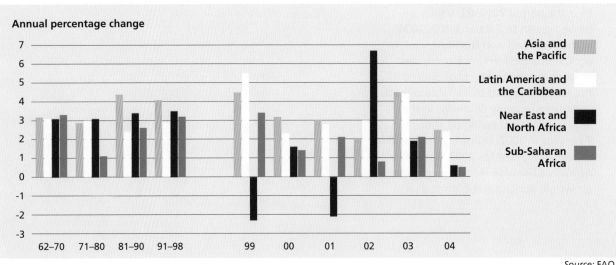

Source: FAO.

FIGURE 45
Long-term trend in per capita food production by region and country group
(Index 1999–2001 = 100)

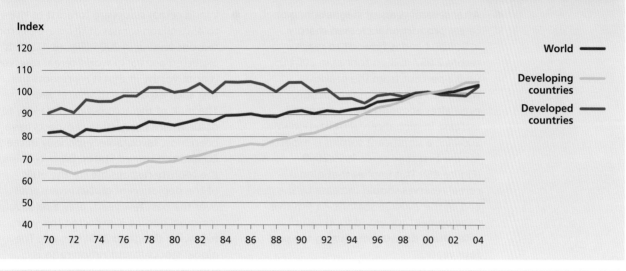

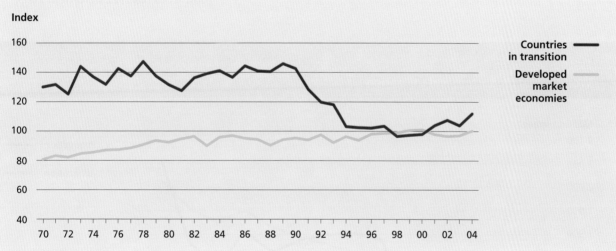

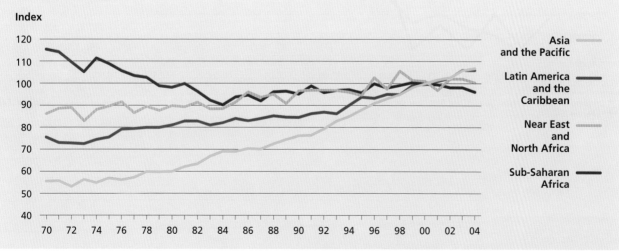

Source: FAO.

5. WORLD CEREAL SUPPLY SITUATION

- After several years of stagnation, global cereal production increased sharply in 2003/04 and is forecast to reach a record 2 057 million tonnes in 2004/05, up 9.2 percent from the previous year. With this level of production, even after allowing for an expected increase in global cereal utilization in 2004/05, a significant surplus is expected for the first time since 1999/2000 (Figure 46). This implies that global cereal reserves should increase by the end of the 2004/05 seasons – a positive development for world food security after sharp drawdowns during the past four years.

- World cereal stocks are forecast to rise to 450 million tonnes at the close of crop seasons ending in 2005 (Figure 47). This expected accrual in world cereal reserves is noteworthy in that it represents the first such expansion in several years. The bulk of the accumulation is likely to occur where production prospects have been most favourable, especially in the EU and the United States. Even in China, the country responsible for the bulk of the depletion of global inventories over the past few years, only a relatively marginal decline is expected this year following the good 2004 harvest. The global stocks-to-utilization ratio is forecast to reach 22 percent in 2005.

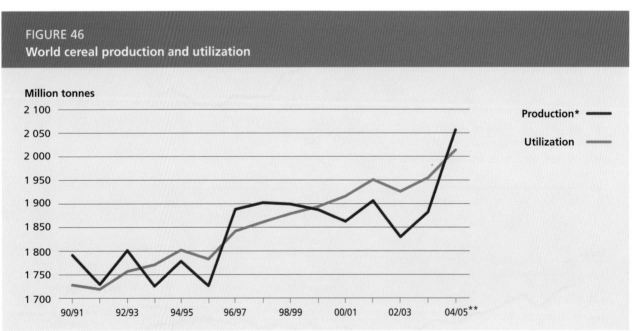

FIGURE 46
World cereal production and utilization

* Data refer to the calendar year of the first year shown.
** Forecast

Source: FAO.

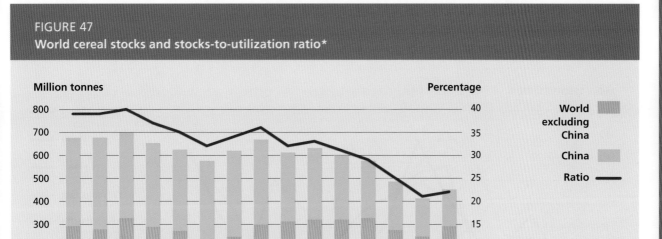

FIGURE 47
World cereal stocks and stocks-to-utilization ratio*

* Stock data are based on aggregate carryovers at the end of national crop years and do not represent world stock levels at any point in time.
** Forecast

Source: FAO.

6. INTERNATIONAL COMMODITY PRICE TRENDS

- In 2004, prices of basic food commodities reached their highest levels since the mid-1990s (Figure 48). Prices of oils and fats have led this trend, rising by 63 percent from the depressed levels of 2000 and 2001. Other basic food prices, including those for cereals, meat and dairy, have also risen, although by smaller margins.

- Price increases in the oilseeds sector reflect continuing strong demand growth for oils for food use and meals for animal feed. The current high level of prices is stimulating farmers to increase plantings, and – assuming weather conditions remain good and pest incidence low – an expected expansion in production in 2004/05 could dampen the upward trend in prices.

- Cereal prices rose by 29 percent between 2000 and 2004. Despite much lower stocks than have prevailed historically, prices moderated somewhat in mid-year on the basis of favourable harvests.

- International meat prices increased in 2003 and 2004 as animal disease outbreaks in major meat-exporting countries and resulting bans on imports from these areas reduced exportable supplies. Poultry and pig meat prices moderated somewhat in 2004, but bovine meat prices continued to surge as disease problems and higher feed prices depress output and trade prospects.

- In contrast with the rising prices of basic food commodities, the price situation for tropical products and raw materials is mixed. The preliminary FAO forecast for the world sugar market in 2005 indicates that world sugar consumption could slightly surpass global production for the second consecutive year. The expected shortfall in global output would lead to falling stocks in major importing countries, underpinning the continued strengthening in market prices.

- Significant oversupply and sluggish demand growth in the world market

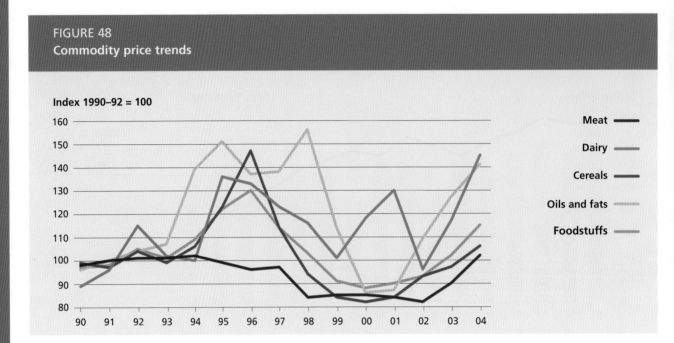

FIGURE 48
Commodity price trends

resulted in coffee prices falling by 58 percent between 1998 and 2001. Prices have remained weak since then and, although some rises occurred in the interim period, it was only in February 2005 that prices actually reached the same level as the 1999 average. Preliminary returns point to a similar crop size in 2004/2005 as that harvested in 2003/04 and a continued upward trend in prices.

- After falling almost by 50 percent between 1998 and 2000, cocoa prices recovered strongly in 2002 and 2003, largely on the basis of disease-reduced harvests. Prices decreased slightly in 2004, but recent difficulties with shipments from West Africa led to a slight strengthening in February 2005.

- Cotton prices declined in late 2004 as a result of record output in the major cotton-producing countries (Brazil, China, India, Pakistan and the United States – which together account for more than 70 percent of world production). World cotton prices were recovering in the first three months of 2005, mostly due to expectations of lower production in 2005/06 following reduced plantings in response to low prices at sowing time.

- Rubber prices also recovered in 2003 and 2004 from the extremely low levels that prevailed during previous years as a result of stronger economic growth and higher prices for petroleum-based synthetic rubber.

- Weak and volatile prices, especially for beverages and other tropical products, have negative effects on the ability of many developing countries to generate export earnings. These effects can be particularly severe for countries that rely on exports of a small number of agricultural commodities for a large share of their export revenues, as do many developing countries. Forty-three developing countries depend on a single agricultural commodity for over 20 percent of their total export revenues and over 50 percent of their agricultural export revenues. Most of these countries are in Latin America and the Caribbean or sub-Saharan Africa. The most important export commodities for these countries are coffee, bananas, cotton lint and cocoa beans. The high dependence on only a few export commodities makes the overall economies of these countries extremely vulnerable to adverse changes in market conditions. Large fluctuations in export proceeds are likely to have negative impacts on income, investment, employment and growth.

FIGURE 48 (cont.)
Commodity price trends

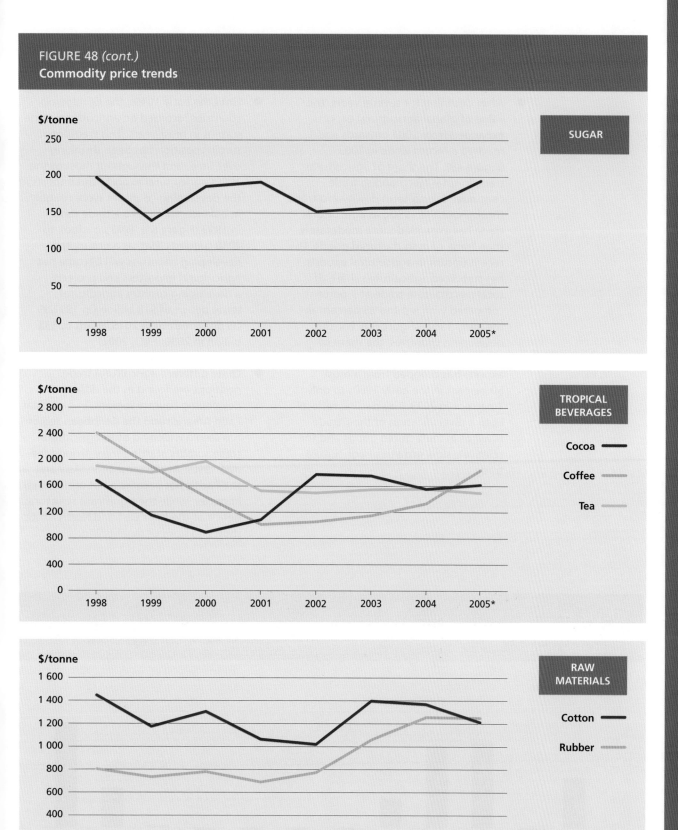

* Data for 2005 are based on a two-month average for coffee, a four-month average for tea and a five-month average for cocoa, rubber and cotton.

Source: FAO.

7. AGRICULTURAL TRADE

- After declining for several years, the value of global agricultural exports expanded from 2001 onwards and reached a record value in 2003 (Figure 49). The share of agricultural trade in total merchandise trade continued a long-term downward trend throughout the 1990s, as agricultural trade has expanded more moderately than trade of manufactured goods. The recent upturn in agricultural exports has stabilized agriculture's share of total merchandise trade at 7 percent, compared with around 25 percent in the early 1960s (Figure 50). For the developing countries, the share of agricultural exports in total merchandise exports has dropped from almost 50 percent in the early 1960s to only 7 percent in 2003. This decline reflects both a diversification of their trade towards manufactured goods and the relatively slow growth of agricultural trade.

- Until the early 1990s, the developing countries recorded an agricultural trade surplus in most years. This traditional surplus position has been shrinking over time, and throughout most of the 1990s agricultural exports and imports in the developing countries were roughly in balance, turning to a trade deficit in 1999 (Figure 51). FAO's outlook to 2030 suggests that, as a group, the developing countries will become net agricultural importers and projects a developing country agricultural trade deficit of $18 billion (in 1997/99 US dollar terms) in 2015, rising to $35 billion in 2030 (FAO, 2002).

- Quite different agricultural trade positions are found in the different developing country regions. In particular, Latin America and the Caribbean region has seen a widening of its agricultural trade surplus, starting from around the mid-1990s. At the same time, Asia and the Pacific has become a net agricultural importer, while the significant structural deficit of the Near East and North Africa shows no signs of diminishing.

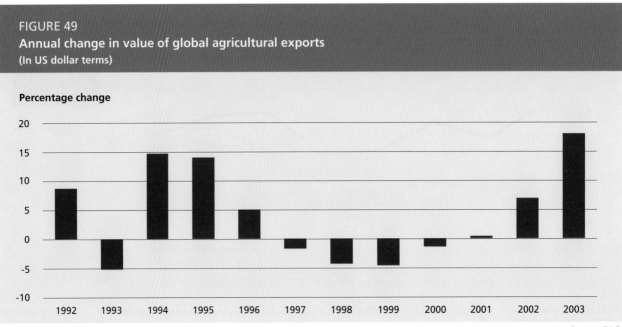

FIGURE 49
Annual change in value of global agricultural exports
(In US dollar terms)

Source: FAO.

FIGURE 50
Global agricultural exports

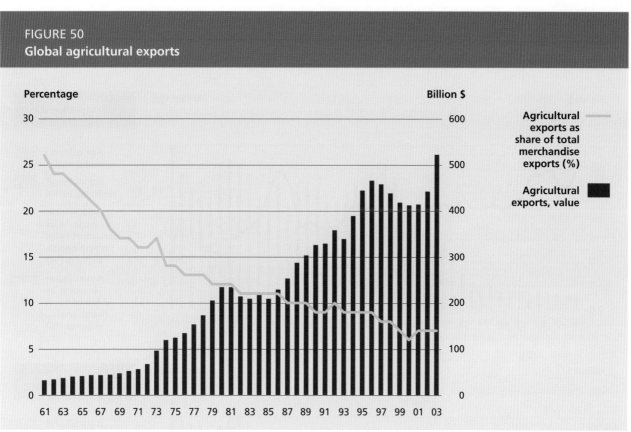

Source: FAO.

FIGURE 51
Agricultural imports and exports, by region

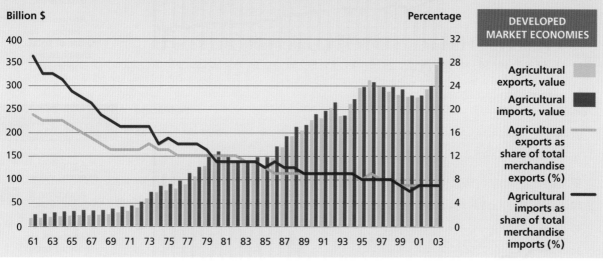

DEVELOPED MARKET ECONOMIES

(Cont.)

FIGURE 51 (cont.)
Agricultural imports and exports, by region

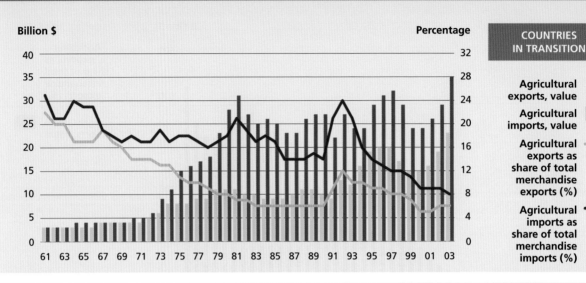

COUNTRIES IN TRANSITION
- Agricultural exports, value
- Agricultural imports, value
- Agricultural exports as share of total merchandise exports (%)
- Agricultural imports as share of total merchandise imports (%)

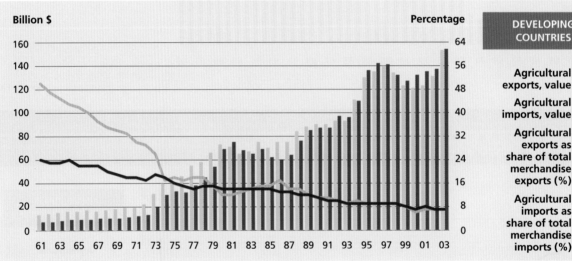

DEVELOPING COUNTRIES
- Agricultural exports, value
- Agricultural imports, value
- Agricultural exports as share of total merchandise exports (%)
- Agricultural imports as share of total merchandise imports (%)

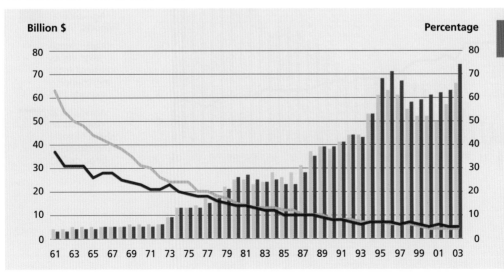

ASIA AND THE PACIFIC
- Agricultural exports, value
- Agricultural imports, value
- Agricultural exports as share of total merchandise exports (%)
- Agricultural imports as share of total merchandise imports (%)

(Cont.)

FIGURE 51 (cont.)
Agricultural imports and exports, by region

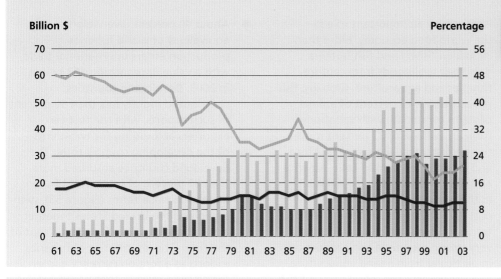

LATIN AMERICA AND THE CARIBBEAN

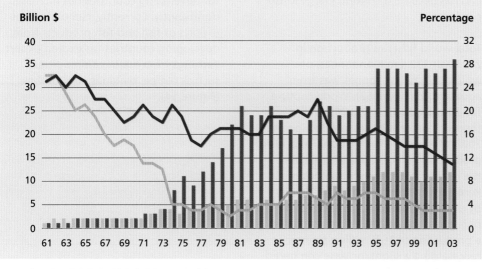

NEAR EAST AND NORTH AFRICA

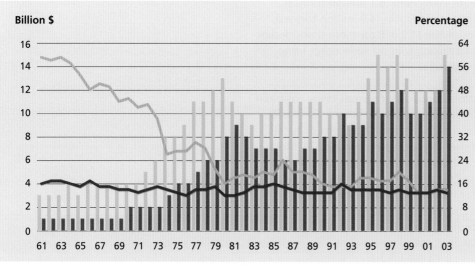

SUB-SAHARAN AFRICA

8. FISHERIES: PRODUCTION, UTILIZATION AND TRADE

- isheries play an important role in the world food economy. More than 38 million fishers and fish farmers gain their livelihoods from capture fisheries and aquaculture. Globally, fish provide about 15 percent of animal proteins consumed, with variations from an average of 23 percent in Asia to approximately 18 percent in Africa and around 7 percent in Latin America and the Caribbean. Developments in the world supply of fish over the past decade have been dominated by trends in China, which has reported very strong growth in fish production, in particular from inland aquaculture, and has become the world's largest fish producer.

- Total world fisheries production in 2003 was 132.5 million tonnes, of which 42.3 million tonnes was from aquaculture (Figure 52). World capture fisheries production was 90.2 million tonnes, 3 percent below production in 2002. Most of the fluctuations in capture production in recent years have been the result of variations in catches of Peruvian anchoveta, which are driven by climatic conditions (i.e. the el Niño phenomenon). In 2003, China reported a production of 16.8 million tonnes, representing a slight increase compared with 2002. Peru (6.1 million tonnes), the United States (4.9 million tonnes), Indonesia (4.7 million tonnes) and Japan (4.6 million tonnes) were other large producers.

- World aquaculture production has been increasing rapidly in recent years and now accounts for 32 percent of total fisheries production (Figure 52). Most of the expansion has been attributable to China, which is now responsible for more than two-thirds of total aquaculture production in volume terms (28.9 million tonnes in 2003).

- About 40 percent (live-weight equivalent) of world fisheries production enters international trade, with a value approaching $63 billion in 2003. Developing countries contributed slightly less than 50 percent of such exports, with the first ten exporters accounting for two-thirds of the developing country total. The developed countries absorbed more than 80 percent of total world fisheries imports in value terms (Figure 53), with Japan and the United States together accounting for as much as 36 percent of the total. The importance of fisheries exports as a foreign currency earner for developing countries has increased significantly. Currently, cumulated net exports of fish and fish products from developing countries far exceed export earnings from major commodities such as coffee, bananas, and rubber (Figure 54).

FIGURE 52
World fish production, China and rest of the world

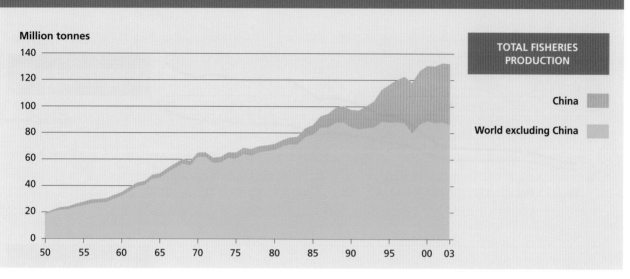

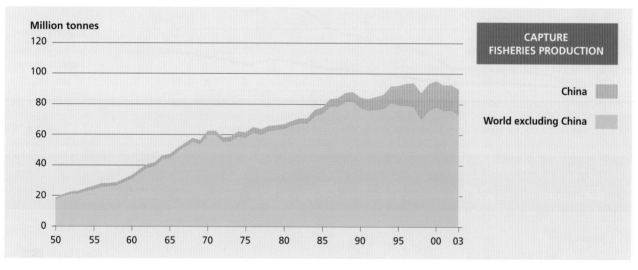

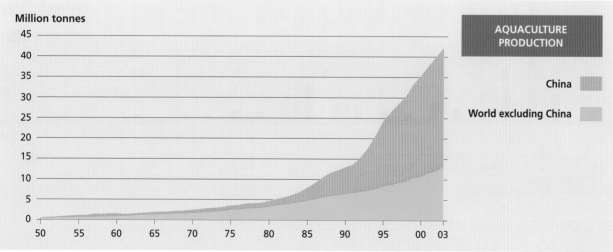

Note: Data exclude production of marine mammals, crocodiles, corals, sponges, shells and aquatic plants.

Source: FAO.

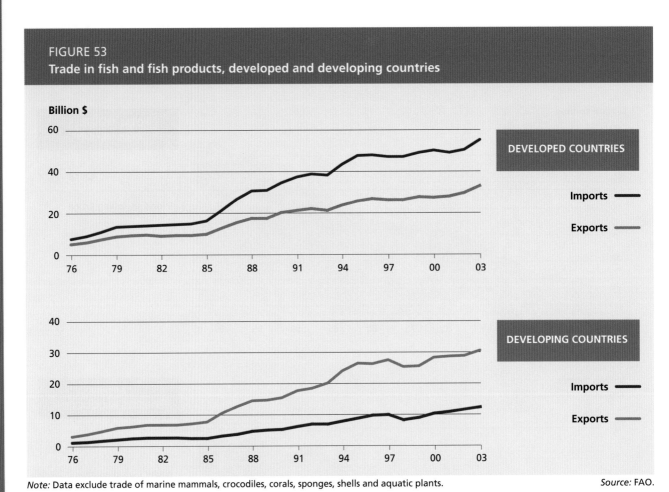

FIGURE 53
Trade in fish and fish products, developed and developing countries

Note: Data exclude trade of marine mammals, crocodiles, corals, sponges, shells and aquatic plants.

Source: FAO.

FIGURE 54
Net exports of fish and fish products and selected agricultural commodities in developing countries

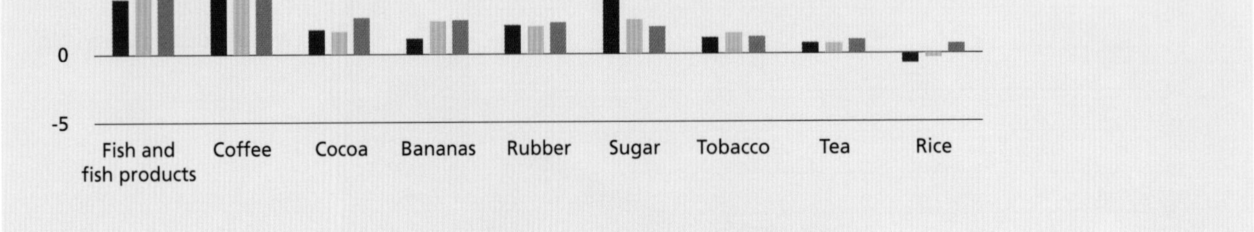

Note: Data exclude trade of marine mammals, crocodiles, corals, sponges, shells and aquatic plants.

Source: FAO.

9. FORESTRY

- World roundwood production in 2003 reached 3 342 million cubic metres, about 1.2 percent above the level of the preceding year (Figure 55). The greater part of global wood production is burned as fuel. Of total roundwood production in 2003, 53 percent was woodfuel and the remaining 47 percent industrial roundwood. The vast majority of wood burning occurs in developing countries, where wood is often the most important source of energy. On the other hand, the larger part of industrial roundwood production continues to be accounted for by the developed countries, which provide more than 70 percent of the total. As most industrial roundwood is consumed and processed domestically, the proportion reaching international markets is small (Figure 56).

- The developing countries accounted for 2 000 million cubic metres, or 60 percent, of total roundwood production in 2002 (Figure 57). Almost 80 percent of roundwood production consists of woodfuel, the production of which has been stable in recent years. Developing country production of industrial roundwood has started to rise slowly after some years of decline. Total roundwood production in the developed countries, following a significant decline in the early 1990s, is still well below the peak levels of 1989–90. Industrial roundwood accounts for 87 percent of production, whereas woodfuel is of relatively marginal importance.

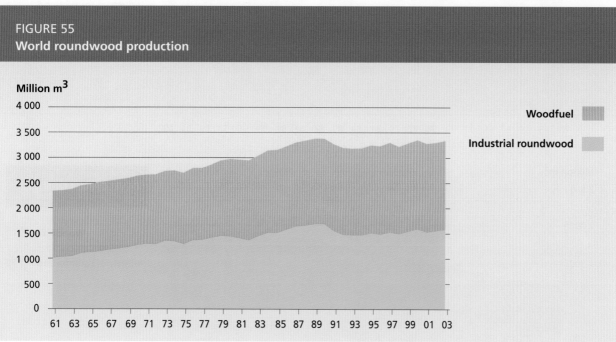

FIGURE 55
World roundwood production

Source: FAO.

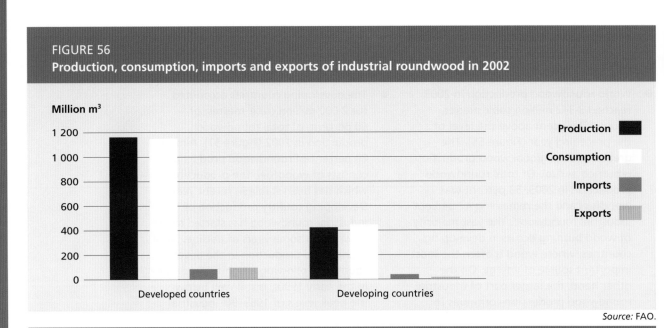

FIGURE 56
Production, consumption, imports and exports of industrial roundwood in 2002

Source: FAO.

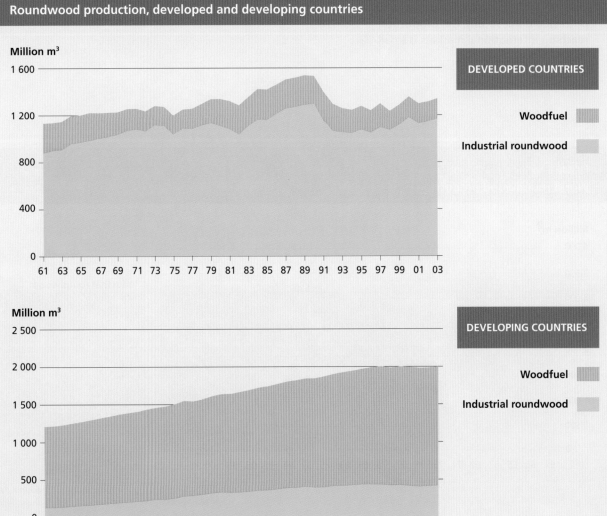

FIGURE 57
Roundwood production, developed and developing countries

Source: FAO.

- The value of international trade in forest products increased rapidly from the mid-1980s to the mid-1990s to reach $155 billion in 2003. Europe, Asia and North and Central America account for the major share of the value of world forest product imports. In 2003, imports of forest products reached a maximum value of $76 billion in Europe and $43 billion in Asia, the second-largest importing region (Figure 58). Exports of forest products increased remarkably in Europe in recent years and topped at $83 billion in 2003, accounting for more than half of the world's total forest product exports. The second-largest exporting region is North and Central America, although trends there are on the decline (Figure 58).

- Europe earns higher trade value both within and outside the region because many countries manufacture value-added products. The region accounts for 55 percent of world export value, although its roundwood production rests at only 30 percent of the world total. Countries in South America, Africa and Oceania mainly trade in raw material, earning 4, 2 and 2 percent, respectively, of world export value while accounting for 10, 4 and 3 percent of total roundwood production (Figure 59).

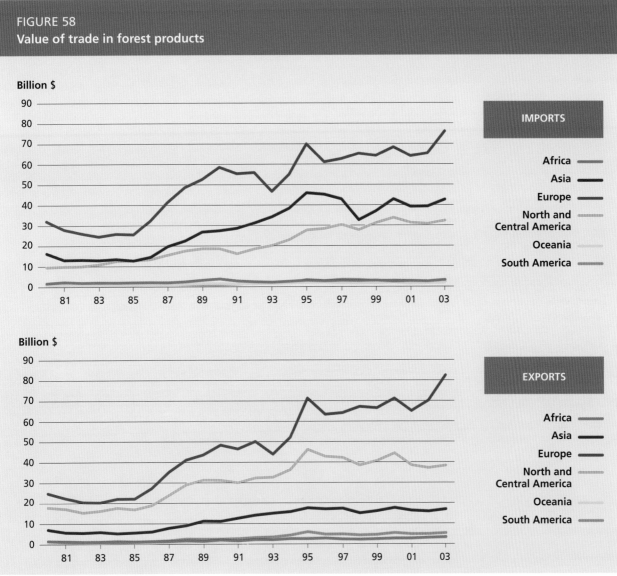

FIGURE 58
Value of trade in forest products

Source: FAO.

FIGURE 59
Industrial roundwood production by region, 2002

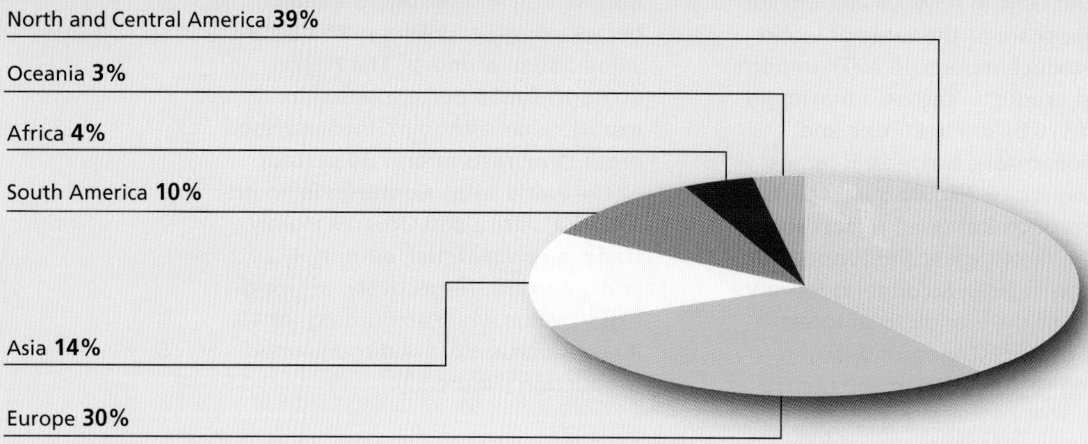

North and Central America **39%**
Oceania **3%**
Africa **4%**
South America **10%**
Asia **14%**
Europe **30%**

Source: FAO.

Part III

STATISTICAL ANNEX

Part III

Notes on the annex tables

Symbols

The following symbols are used in the tables:

...	= not available
ha	= hectare
hg/ha	= hectogram per hectare
hg	= hectogram
GDP	= gross domestic product
GNP	= gross national product
kcal/person/day	= calories per person per day
kg	= kilogram
US$	= US dollar

To divide decimals from whole number a full point (.) is used.

Technical notes

The tables do not include countries for which there were insufficient data.

Numbers displayed in the tables might be slightly different from those obtained from FAOSTAT and the World Development Indicators because of rounding.

1. Food security and nutrition (Table A2)
Source: FAO

Undernourishment
FAO's estimates of the prevalence of undernourishment are based on calculations of the amount of food available in each country (national dietary energy supply or DES) and a measure of inequality in distribution derived from household income or expenditure surveys.

Although not listed separately, provisional estimates for Afghanistan, Iraq, Papua New Guinea and Somalia have been included in the relevant regional aggregates.

Eritrea and Ethiopia were not separate entities in 1990–1992, but estimates of the number and proportion of undernourished in the former Ethiopia PDR are included in regional and subregional aggregates for that period.

Symbols used
To denote a proportion of less than 2.5 percent undernourished a dash (–) is used.

Dietary energy supply
Per capita supplies in terms of product weight are derived from the total supplies available for human consumption (i.e. food) by dividing the quantities of food by the total population actually

partaking of the food supplies during the reference period. Dietary energy supply is weighted by the total population.

2. Agricultural production and productivity (Table A3)
Source: FAO

Agricultural and per capita food production annual growth rates
The growth rates refer to the level of change of the aggregate volume of production. Production quantities of each commodity are weighted by 1999–2001 average international commodity prices and summed for each year.

3. Population and labour force indicators (Table A4)
Source: FAO

Total population
The total population usually refers to the present-in-area (de facto) population, which includes all persons physically present within the current geographical boundaries of countries at the mid-point of the reference period.

Rural population
Usually the urban area is defined and the residual from the total population is taken as rural. In practice, the criteria adopted for distinguishing between urban and rural areas vary among countries.

Agricultural population
The agricultural population is defined as all persons depending for their livelihood on agriculture, hunting, fishing or forestry. This estimate comprises all persons actively engaged in agriculture and their non-working dependants.

Economically active population
This refers to the number of all employed and unemployed persons (including those seeking work for the first time).

Economically active population in agriculture
The economically active population in agriculture is that part of the economically active population engaged in or seeking work in agriculture, hunting, fishing or forestry.

4. Land-use indicators (Table A5)
Source: FAO

Total land area
Total area excluding area under inland water bodies.

Forest and wood area
Land under natural or planted stands of trees, whether productive or not.

Agricultural area
The sum of area under arable land, permanent crops and permanent pastures.

Arable land
Land under temporary crops (double-cropped areas are counted only once), temporary meadows for mowing or pasture, land under market- and kitchen-gardens and land temporarily fallow (less than five years).

Permanent crops area
Land cultivated with crops that occupy the land for long periods and need not be replanted after each harvest.

Permanent pasture area
Land used permanently (five years or more) for herbaceous forage crops, either cultivated or growing wild (wild prairie or grazing land).

Irrigated area
Data on irrigation relate to areas equipped to provide water to the crops.
- *China*: data on irrigated area cover farmland only (areas under orchard and pastures are excluded).
- *Cuba*: data refer to state sector only.
- *Japan; Republic of Korea; Sri Lanka*: data refer to irrigated rice only.

Fertilizer consumption (use)
Data refer to total fertilizer use. The total estimates are obtained by adding the volumes of nitrogenous, phosphate and potash fertilizers expressed in terms of plant nutrients (N, P_2O_5 and K_2O, respectively).

5. Trade indicators (Table A6)
Source: FAO and World Bank (*World Development Indicators 2005*, CD-ROM and online dataset)

Total merchandise trade
Data refer to the total merchandise trade. In general, export values are f.o.b. (free on board) and import values are c.i.f. (cost, insurance and freight).

Agricultural trade
Data refer to agriculture in the narrow sense, excluding fishery and forestry products.

Food trade
Data refer to food and animals.

Agricultural GDP
The agriculture, value added (percentage of GDP), is derived from World Bank national accounts data, and OECD National Account data files. Agriculture includes forestry, fishing and hunting, as well as cultivation of crops and livestock production.

Agricultural exports relative to agricultural GDP
Agricultural exports relative to agricultural GDP was weighted by agriculture, value added.

6. Economic indicators (Table A7)
Source: World Bank (*World Development Indicators 2005*, CD-ROM and online dataset)

Weighting: GNP per capita (current US$), GDP per capita (annual percentage growth) and GDP per capita, PPP (current international $) were weighted by the total population. GDP (annual percentage growth) and agriculture, value added (percentage of GDP), were weighted by GDP (constant 2000 US$). Agriculture, value added (annual percentage growth), was weighted by agriculture, value added (constant 2000 US$). Agriculture, value added per worker, was weighted by economic active population in agriculture.

National poverty headcount
National poverty rate is the percentage of the population living below the national poverty line. National estimates are based on population-weighted subgroup estimates from household surveys.

GNP per capita (current US$)
GNP per capita is the gross national income, converted to US dollars using the World Bank Atlas method, divided by the mid-year population.

GDP (annual percentage growth)
Annual percentage growth rate of GDP at market prices based on constant local currency. Aggregates are based on constant 2000 US dollars.

GDP per capita (annual percentage growth)
Annual percentage growth rate of GDP per capita based on constant local currency. GDP per capita is the GDP divided by mid-year population.

GDP per capita, PPP (current international $)
GDP per capita based on purchasing power parity (PPP). PPP GDP is gross domestic product converted to international dollars using purchasing power parity rates. An international dollar has the same purchasing power over GDP as the US dollar has in the United States of America.

Agriculture, value added per worker
Agriculture, value added per worker, is a measure of agricultural productivity. Value added in agriculture measures the output of the agriculture sector less the value of intermediate inputs. Agriculture comprises value added from forestry, hunting and fishing as well as cultivation of crops and livestock production.

GDP, constant 2000 US$
Data are in constant 2000 US dollars. Dollar figures for GDP are converted from domestic currencies using 2000 official exchange rates.

7. Total factor productivity (Table A8)
Source: FAO

Total factor productivity (TFP) measures the quantity of output divided by a measure of the quantity of inputs used. The approach taken here is to apply data envelopment analysis (DEA) methods to output and input data obtained from FAOSTAT to estimate a Malmquist index of TFP (Malmquist, 1953). The data cover the periods 1961–80 and 1981–2000. The resulting change in total productivity index can be disaggregated into a technology component and a technical efficiency component. A distinct advantage of the Malmquist DEA method is that no information on input prices is required. The data used are as follows: Output is net agricultural production, i.e. excluding seed and feed, in constant (1989–91) "international dollars"; Inputs are: Land: arable and land under permanent crops; Labour: total population economically active in agriculture; Fertilizer: total consumption (in nutrient-equivalent terms) of nitrogen, potash and phosphates; Livestock: the weighted sum of camels, buffalo, horses, cattle, asses, pigs, sheep, goats and poultry (using the weights suggested by Hayami and Ruttan, 1985); Physical capita: number of tractors in use. We also included the proportion of arable and permanent cropland that is irrigated as well as the ratio of land that is arable and under permanent crops to agricultural area (which also includes permanent pastures).

Data for Ethiopia and Eritrea start in 1993 instead of 1981.

Country and regional notes

Data for China do not include data for Hong Kong Special Administrative Region; Macao Special Administrative Region or Taiwan Province of China, unless otherwise noted.
Data are shown for Belgium and Luxembourg separately whenever possible, but in most cases before 2000 the data are aggregated in Belgium/Luxembourg.

Data are shown whenever possible for the individual countries formed from the Ethiopia PDR – Eritrea and Ethiopia. Data for the years prior to 1992 are shown under Ethiopia PDR.

Data for Yemen refer to that country from 1990 onward; data for previous years refer to aggregated data of the former People's Democratic Republic of Yemen and the former Yemen Arab Republic unless otherwise noted.

South Africa is included in sub-Saharan Africa and not in the developed countries.

TABLE A1
Countries and territories used for statistical purposes in this publication

Developing countries				Developed countries	
Asia and the Pacific/ Far East and Oceania	Latin America and the Caribbean	Near East and North Africa	Sub-Saharan Africa	Developed market economies	Countries in transition
American Samoa	Anguilla	Afghanistan	Angola	Andorra	Albania
Bangladesh	Antigua and Barbuda	Algeria	Benin	Australia	Armenia
Bhutan	Argentina	Bahrain	Botswana	Austria	Azerbaijan
British Virgin Islands	Aruba	Cyprus	Burkina Faso	Belgium-Luxembourg	Belarus
Brunei Darussalam	Bahamas	Egypt	Burundi	Canada	Bosnia and Herzegovina
Cambodia	Barbados	Iran, Islamic Rep. of	Cameroon	Denmark	Bulgaria
China, Hong Kong SAR	Belize	Iraq	Cape Verde	Faeroe Islands	Croatia
China, Macao SAR	Bermuda	Jordan	Central African Republic	Finland	Czech Republic
China, Mainland	Bolivia	Kuwait	Chad	France	Estonia
China, Taiwan Province of	Brazil	Lebanon	Comoros	Germany	Georgia
Cocos (Keeling) Islands	Cayman Islands	Libyan Arab Jamahiriya	Congo	Gibraltar	Hungary
Cook Islands	Chile	Morocco	Congo, Democratic Republic of the	Greece	Kazakhstan
Fiji	Colombia	Occupied Palestinian Territory	Côte d'Ivoire	Greenland	Kyrgyzstan
French Polynesia	Costa Rica	Oman	Djibouti	Iceland	Latvia
Guam	Cuba	Qatar	Equatorial Guinea	Ireland	Lithuania
India	Dominica	Saudi Arabia	Eritrea	Israel	Macedonia, The former Yugoslav Republic of
Indonesia	Dominican Republic	Syrian Arab Republic	Ethiopia	Italy	Moldova, Republic of
Kiribati	Ecuador	Tunisia	Gabon	Japan	Poland
Korea, Dem. People's Republic of	El Salvador	Turkey	Gambia	Liechtenstein	Romania
Korea, Republic of	Falkland Islands (Malvinas)	United Arab Emirates	Ghana	Malta	Russian Federation
Lao People's Democratic Republic	French Guiana	Yemen	Guinea	Monaco	Serbia and Montenegro
Malaysia	Grenada		Guinea-Bissau	Netherlands	Slovakia
Maldives	Guadeloupe		Kenya	New Zealand	Slovenia
Marshall Islands	Guatemala		Lesotho	Norway	Tajikistan
Micronesia, Fed. States of	Guyana		Liberia	Portugal	Turkmenistan
Mongolia	Haiti		Madagascar	Saint Pierre and Miquelon	Ukraine
Myanmar	Honduras		Malawi	San Marino	Uzbekistan
Nauru	Jamaica		Mali	Spain	
Nepal	Martinique		Mauritania	Sweden	
New Caledonia	Mexico		Mauritius	Switzerland	
Niue	Montserrat		Mozambique	United Kingdom	
Norfolk Island	Netherlands Antilles		Namibia	United States of America	
Northern Marianas Islands	Nicaragua		Niger		
Pakistan	Panama		Nigeria		
Palau	Paraguay		Réunion		
Papua New Guinea	Peru		Rwanda		
Philippines	Puerto Rico		Saint Helena		

TABLE A1 (cont.)

Developing countries				Developed countries	
Asia and the Pacific/ Far East and Oceania	**Latin America and the Caribbean**	**Near East and North Africa**	**Sub-Saharan Africa**	**Developed market economies**	**Countries in transition**
Samoa	Saint Kitts and Nevis		Sao Tome and Principe		
Singapore	Saint Lucia		Senegal		
Solomon Islands	Saint Vincent and the Grenadines		Seychelles		
Sri Lanka	Suriname		Sierra Leone		
Thailand	Trinidad and Tobago		Somalia		
Timor-Leste	Turks and Caicos Islands		South Africa		
Tokelau	United States Virgin Islands		Sudan		
Tonga	Uruguay		Swaziland		
Tuvalu	Venezuela, Bolivarian Republic of		Tanzania, United Rep. of		
Vanuatu			Togo		
Viet Nam			Uganda		
Wallis and Futuna Islands			Zambia		
			Zimbabwe		

TABLE A2
Food security and nutrition

	Number of people undernourished (Millions)		Proportion of undernourished in total population (%)		Dietary energy supply (kcal/person/day)		(Average annual % increase)
	1990–92	2000–02	1990–92	2000–02	1990–92	2000–02	1990–2002
WORLD	...	...	...	...	2 708	2 803	0.28
DEVELOPED COUNTRIES	...	...	...	...	3 273	3 314	0.01
DEVELOPING COUNTRIES	823.8	814.6	20	17	2 537	2 667	0.47
ASIA AND THE PACIFIC	569.2	519	20	16	2 521	2 674	0.53
Bangladesh	39.2	42.5	35	30	2 070	2 190	0.60
Brunei Darussalam	...	...	...	...	2 797	2 855	0.21
Cambodia	4.3	4.4	43	33	1 871	2 059	1.18
China, Hong Kong SAR	...	...	...	...	3 239	3 115	–0.42
China, Macao SAR	...	...	...	...	2 725	2 498	–0.41
China, Mainland	193.5	142.1	16	11	2 699	2 957	0.88
China, Taiwan Province of	...	...	...	...	2 964	2 997	0.22
Fiji	...	...	...	...	2 637	2 895	0.95
French Polynesia	...	...	...	...	2 864	2 884	0.12
India	215.8	221.1	25	21	2 366	2 420	0.19
Indonesia	16.4	12.6	9	6	2 698	2 912	0.82
Kiribati	...	...	...	...	2 653	2 852	0.97
Korea, Democratic People's Rep. of	3.7	8.1	18	36	2 452	2 138	–0.26
Korea, Republic of	0.8	0.7	–	–	2 999	3 059	–0.03
Lao People's Dem. Rep.	1.2	1.2	29	22	2 111	2 285	0.73
Malaysia	0.5	0.6	3	–	2 822	2 891	0.46
Maldives	...	...	...	...	2 377	2 542	0.69
Mongolia	0.8	0.7	34	28	2 065	2 236	0.25
Myanmar	4	2.8	10	6	2 634	2 880	0.95
Nepal	3.9	4	20	17	2 346	2 443	0.24
New Caledonia	...	...	...	...	2 792	2 790	–0.02
Pakistan	27.7	29.3	24	20	2 305	2 431	0.17
Philippines	16.2	17.2	26	22	2 264	2 375	0.28
Samoa	...	...	...	...	2 569	2 900	0.80
Solomon Islands	...	...	...	...	2 015	2 238	0.49
Sri Lanka	4.8	4.1	28	22	2 229	2 388	0.44
Thailand	15.2	12.2	28	20	2 252	2 453	0.70
Timor-Leste	...	...	...	...	2 564	2 813	1.13
Vanuatu	...	...	...	...	2 524	2 572	0.10
Viet Nam	20.6	14.7	31	19	2 177	2 534	1.48
LATIN AMERICA AND THE CARIBBEAN	59.5	52.9	13	10	2 705	2 848	0.49
Antigua and Barbuda	...	...	...	...	2 460	2 342	–0.22
Argentina	0.7	0.6	–	–	2 993	3 075	0.06
Bahamas	...	...	...	...	2 619	2 753	–0.08
Barbados	...	...	...	...	3 080	3 059	–0.14
Belize	...	...	...	...	2 651	2 844	0.92
Bermuda	...	...	...	...	2 341	2 257	–0.32
Bolivia	1.9	1.8	28	21	2 112	2 250	0.41
Brazil	18.5	15.6	12	9	2 812	3 010	0.76

TABLE A2 (cont.)

	Number of people undernourished (Millions)		Proportion of undernourished in total population (%)		Dietary energy supply (kcal/person/day)		(Average annual % increase)
	1990–92	2000–02	1990–92	2000–02	1990–92	2000–02	1990–2002
Chile	1.1	0.6	8	4	2 611	2 845	1.13
Colombia	6.1	5.7	17	13	2 435	2 579	0.66
Costa Rica	0.2	0.2	6	4	2 714	2 858	0.44
Cuba	0.8	0.4	8	3	2 717	2 998	0.50
Dominica	...	...	...	...	2 941	2 752	–0.60
Dominican Republic	1.9	2.1	27	25	2 261	2 323	0.10
Ecuador	0.9	0.6	8	4	2 509	2 737	0.79
El Salvador	0.6	0.7	12	11	2 492	2 548	0.58
Grenada	...	...	...	...	2 830	2 867	0.42
Guatemala	1.4	2.8	16	24	2 351	2 187	–0.11
Guyana	0.2	0.1	21	9	2 347	2 709	0.91
Haiti	4.6	3.8	65	47	1 780	2 083	1.46
Honduras	1.1	1.5	23	22	2 313	2 353	0.17
Jamaica	0.3	0.3	14	10	2 503	2 675	0.26
Mexico	4.6	5.2	5	5	3 101	3 155	0.17
Netherlands Antilles	...	...	...	...	2 523	2 557	0.10
Nicaragua	1.2	1.4	30	27	2 216	2 283	0.14
Panama	0.5	0.8	21	26	2 316	2 237	0.24
Paraguay	0.8	0.8	18	14	2 403	2 556	0.00
Peru	9.3	3.4	42	13	1 962	2 550	1.38
Saint Kitts and Nevis	...	...	...	...	2 577	2 636	–0.16
Saint Lucia	...	...	...	...	2 735	2 936	0.95
Saint Vincent and the Grenadines	...	...	...	...	2 299	2 534	1.04
Suriname	0.1	0	13	11	2 528	2 628	0.73
Trinidad and Tobago	0.2	0.2	13	12	2 635	2 734	0.14
Uruguay	0.2	0.1	6	4	2 661	2 828	0.97
Venezuela, Bolivarian Republic of	2.3	4.3	11	17	2 464	2 351	0.06
NEAR EAST AND NORTH AFRICA	**24.8**	**39.2**	**8**	**10**	**3 070**	**3 106**	**0.17**
Algeria	1.3	1.7	5	5	2 921	2 991	0.36
Cyprus	...	...	...	...	3 122	3 251	0.65
Egypt	2.5	2.4	4	3	3 200	3 341	0.50
Iran, Islamic Republic of	2.1	2.7	4	4	2 978	3 075	0.49
Jordan	0.1	0.4	4	7	2 818	2 668	–0.13
Kuwait	0.5	0.1	23	5	2 292	3 052	0.67
Lebanon	0.1	0.1	–	3	3 165	3 181	0.18
Libyan Arab Jamahiriya	0	0	–	–	3 277	3 324	0.15
Morocco	1.5	2	6	7	3 029	3 042	0.17
Saudi Arabia	0.7	0.8	4	3	2 772	2 845	0.39
Syrian Arab Republic	0.7	0.6	5	4	2 831	3 038	0.68
Tunisia	0.1	0.1	–	–	3 152	3 271	0.31
Turkey	1	1.8	–	3	3 494	3 359	–0.31
United Arab Emirates	0.1	0.1	4	–	2 928	3 199	0.56
Yemen	4.2	6.7	34	36	2 036	2 037	–0.43
SUB–SAHARAN AFRICA	**170.4**	**203.5**	**36**	**33**	**2 175**	**2 254**	**0.40**
Angola	5.6	5.1	58	40	1 783	2 041	1.33
Benin	1	0.9	20	15	2 338	2 516	0.83
Botswana	0.3	0.6	23	32	2 263	2 155	–0.06
Burkina Faso	1.9	2.3	21	19	2 353	2 408	0.79

TABLE A2 (cont.)

	Number of people undernourished (Millions)		Proportion of undernourished in total population (%)		Dietary energy supply		
					(kcal/person/day)		(Average annual % increase)
	1990–92	2000–02	1990–92	2000–02	1990–92	2000–02	1990–2002
Burundi	2.7	4.4	48	68	1 896	1 636	–0.53
Cameroon	4	3.9	33	25	2 114	2 266	0.64
Cape Verde	...	...	...	...	3 011	3 209	0.86
Central African Republic	1.5	1.6	50	43	1 874	1 977	0.43
Chad	3.5	2.7	58	34	1 783	2 146	1.84
Comoros	...	...	...	...	1 914	1 748	–0.48
Congo	1.4	1.3	54	37	1 861	2 086	0.72
Congo, Democratic Republic of the	12.2	35.5	32	71	2 173	1 627	–2.37
Côte d'Ivoire	2.3	2.2	18	14	2 472	2 621	0.53
Djibouti	...	...	...	...	1 802	2 202	1.71
Eritrea	...	2.8	...	73	...	1 524	...
Ethiopia	...	31.3	...	46	...	1 843	...
Ethiopia PDR	...	...	...	...	1 638	...	...
Gabon	0.1	0.1	10	6	2 455	2 614	0.48
Gambia	0.2	0.4	22	27	2 367	2 269	–0.22
Ghana	5.8	2.5	37	13	2 073	2 619	2.44
Guinea	2.5	2.1	39	26	2 105	2 382	1.55
Guinea–Bissau	...	...	...	...	2 300	2 101	–0.65
Kenya	10.7	10.3	44	33	1 921	2 107	0.29
Lesotho	0.3	0.2	17	12	2 445	2 617	0.76
Liberia	0.7	1.4	34	46	2 211	1 997	–2.05
Madagascar	4.3	6	35	37	2 084	2 061	–0.43
Malawi	4.8	3.8	50	33	1 881	2 155	0.85
Mali	2.7	3.6	29	29	2 216	2 200	–0.27
Mauritania	0.3	0.3	15	10	2 556	2 771	0.70
Mauritius	0.1	0.1	6	6	2 887	2 955	0.53
Mozambique	9.2	8.5	66	47	1 735	2 033	1.34
Namibia	0.5	0.4	35	22	2 061	2 269	0.82
Niger	3.2	3.8	41	34	2 020	2 130	0.24
Nigeria	11.8	11	13	9	2 538	2 705	1.31
Rwanda	2.8	3	44	37	1 947	2 049	0.49
Sao Tome and Principe	...	...	...	...	2 272	2 390	0.68
Senegal	1.8	2.3	23	24	2 276	2 280	0.49
Seychelles	...	...	...	...	2 311	2 453	0.47
Sierra Leone	1.9	2.3	46	50	1 991	1 926	0.03
South Africa	...	...	...	...	2 827	2 917	0.53
Sudan	8	8.5	32	27	2 159	2 260	0.29
Swaziland	0.1	0.2	14	19	2 455	2 360	–0.40
Tanzania, United Republic of	9.9	15.6	37	44	2 050	1 959	–0.75
Togo	1.2	1.2	33	26	2 151	2 296	0.70
Uganda	4.2	4.6	24	19	2 275	2 363	0.23
Zambia	4	5.2	48	49	1 929	1 904	–0.25
Zimbabwe	4.9	5.6	45	44	1 975	2 024	–0.61
DEVELOPED MARKET ECONOMIES	...	...	...	...	3 332	3 491	0.42
Australia	...	...	...	...	3 177	3 090	–0.35
Austria	...	...	...	...	3 512	3 738	0.48
Belgium–Luxembourg	...	...	...	...	3 581	...	...
Belgium	...	...	...	...	...	3 583	...
Canada	...	...	...	...	3 056	3 560	1.31

TABLE A2 (cont.)

	Number of people undernourished (Millions)		Proportion of undernourished in total population (%)		Dietary energy supply (kcal/person/day)		(Average annual % increase)
	1990–92	2000–02	1990–92	2000–02	1990–92	2000–02	1990–2002
Denmark	...	...	...	...	3 232	3 409	0.57
Finland	...	...	...	...	3 185	3 122	–0.34
France	...	...	...	...	3 535	3 634	0.20
Germany	...	...	...	...	3 397	3 474	0.13
Greece	...	...	...	...	3 566	3 688	0.22
Iceland	...	...	...	...	3 095	3 222	0.26
Ireland	...	...	...	...	3 632	3 664	0.10
Israel	...	...	...	...	3 390	3 644	0.58
Italy	...	...	...	...	3 591	3 690	0.25
Japan	...	...	...	...	2 813	2 783	–0.22
Luxembourg	...	...	...	...	...	3 590	...
Malta	...	...	...	...	3 240	3 539	0.67
Netherlands	...	...	...	...	3 350	3 354	0.44
New Zealand	...	...	...	...	3 215	3 220	0.24
Norway	...	...	...	...	3 181	3 425	0.72
Portugal	...	...	...	...	3 449	3 749	0.88
Spain	...	...	...	...	3 305	3 363	0.39
Sweden	...	...	...	...	2 990	3 135	0.53
Switzerland	...	...	...	...	3 307	3 472	0.47
United Kingdom	...	...	...	...	3 265	3 397	0.40
United States of America	...	...	...	...	3 502	3 795	0.78
	1993–95	2000–02	1993–95	2000–02	1993–95	2000–02	1993–2002
COUNTRIES IN TRANSITION	23.3	28.3	6	7	2 950	2 939	–0.04
Albania	0.2	0.2	5	6	2 881	2 861	1.14
Armenia	1.8	1.1	52	34	1 957	2 190	2.87
Azerbaijan	2.6	1.2	34	15	2 134	2 481	1.31
Belarus	0.1	0.2	–	–	3 177	3 006	–0.52
Bosnia and Herzegovina	0.3	0.3	9	8	2 685	2 762	1.61
Bulgaria	0.7	0.8	8	11	2 894	2 801	–0.89
Croatia	0.7	0.3	16	7	2 517	2 771	1.45
Czech Republic	0.2	0.2	–	–	3 076	3 118	...
Estonia	0.1	0.1	9	5	2 749	2 993	1.64
Georgia	2.1	1.4	39	27	2 106	2 276	1.12
Hungary	0.1	0	–	–	3 344	3 471	–0.33
Kazakhstan	0.2	2	–	13	3 274	2 546	–0.80
Kyrgyzstan	0.9	0.3	21	6	2 410	2 950	1.35
Latvia	0.1	0.1	3	4	2 966	2 958	0.15
Lithuania	0.2	0	4	–	2 874	3 360	1.19
Macedonia, The former Yugoslav Republic of	0.3	0.2	15	11	2 512	2 639	0.69
Moldova, Republic of	0.2	0.5	5	11	2 929	2 717	–1.32
Poland	0.3	0.3	–	–	3 336	3 376	0.07
Romania	0.4	0.2	–	–	3 210	3 414	1.59
Russian Federation	6.4	5.2	4	4	2 926	3 000	0.51
Serbia and Montenegro	0.5	1.1	5	11	2 900	2 660	–1.32
Slovakia	0.2	0.3	4	5	2 928	2 877	...
Slovenia	0.1	0	3	–	2 945	3 015	0.88
Tajikistan	1.2	3.7	21	61	2 314	1 836	–2.36
Turkmenistan	0.5	0.4	13	9	2 509	2 719	–0.07
Ukraine	1.2	1.5	–	3	3 029	2 985	–0.90
Uzbekistan	1.7	6.6	8	26	2 649	2 270	–1.62

TABLE A3
Agricultural production and productivity

	Crop and livestock production		Per capita food production		Cereal yields	
	(Average annual rate of growth [%])				(hg/ha)	
	1985–1994	1995–2004	1985–1994	1995–2004	1992–1994	2002–2004
WORLD	1.9	2.5	0.3	1.2	28 002	31 675
DEVELOPED COUNTRIES	–0.1	1.0	–0.7	0.6	32 087	38 038
DEVELOPING COUNTRIES	3.4	3.3	1.6	1.8	25 518	28 363
ASIA AND THE PACIFIC	3.7	3.6	2.1	2.3	30 889	34 590
Bangladesh	1.6	3.7	–0.8	1.7	25 831	34 098
Bhutan	1.0	–0.5	–1.2	–3.1	12 269	15 908
Brunei Darussalam	2.5	10.4	–0.4	7.8	17 202	16 667
Cambodia	5.9	5.7	2.0	3.2	13 661	20 416
China, Mainland	4.4	4.8	3.5	4.0	44 763	49 802
China, Taiwan Province of	1.1	–0.4	0.1	–1.0	53 957	60 183
Cocos (Keeling) Islands	1.4	2.5	...	...	...	...
Fiji	0.8	–0.8	0.0	–1.9	22 434	22 911
French Polynesia	2.7	0.7	0.5	–0.9	...	...
Guam	1.4	2.3	–0.6	0.9	20 000	20 000
India	3.2	2.3	1.2	0.6	20 751	23 129
Indonesia	3.7	2.5	2.0	1.1	38 663	42 294
Kiribati	2.6	2.6	0.5	1.1	...	...
Korea, Dem. People's Rep. of	2.2	0.1	0.7	–0.6	54 867	33 103
Korea, Republic of	2.3	0.8	1.3	0.2	58 524	59 909
Lao People's Dem. Rep.	3.8	5.9	1.1	3.5	24 869	31 741
Malaysia	4.9	3.4	3.9	1.5	29 960	32 749
Maldives	2.7	3.3	–0.4	0.2	10 000	10 000
Micronesia	0.0	...	–0.3	...	...	...
Mongolia	–1.4	2.4	–3.9	1.5	8 124	6 757
Myanmar	1.7	4.3	0.1	2.8	28 726	35 171
Nepal	2.9	3.3	0.6	1.0	17 860	22 575
New Caledonia	2.0	0.7	0.0	–1.4	28 134	36 598
Pakistan	4.6	3.1	1.8	0.6	18 928	23 322
Papua New Guinea	1.9	2.2	–0.7	–0.1	27 594	37 606
Philippines	2.4	3.0	0.2	1.1	22 095	28 251
Samoa	–2.7	2.0	–3.1	1.1	...	...
Singapore	–14.5	–5.9	–16.5	–8.1	...	...
Solomon Islands	0.0	3.4	–3.2	0.3	0	38 737
Sri Lanka	1.0	0.2	–0.3	–0.9	29 613	33 052
Thailand	2.5	1.3	0.7	0.0	22 982	27 079
Timor-Leste	3.6	0.6	1.2	0.3	19 308	19 941
Tonga	1.0	–0.7	0.8	–1.3	...	...
Vanuatu	–0.4	0.2	–2.9	–2.4	5 205	5 385
Viet Nam	4.2	5.3	1.8	3.5	33 443	44 844
LATIN AMERICA AND THE CARIBBEAN	2.6	3.2	0.9	1.7	24 563	30 121
Antigua and Barbuda	2.5	0.4	2.2	–0.6	18 333	15 709
Argentina	1.5	2.4	0.3	1.2	29 066	32 119
Bahamas	–0.1	5.9	–2.0	4.5	16 866	20 609

TABLE A3 (cont.)

	Crop and livestock production		Per capita food production		Cereal yields	
	(Average annual rate of growth [%])				(hg/ha)	
	1985–1994	1995–2004	1985–1994	1995–2004	1992–1994	2002–2004
Barbados	–1.8	1.5	–2.1	1.2	26 127	26 093
Belize	4.8	4.3	2.1	1.9	19 944	27 603
Bolivia	4.4	3.7	2.1	1.6	14 658	18 796
Brazil	3.8	4.5	2.3	3.0	22 606	31 292
Chile	5.0	1.9	3.4	0.6	43 042	52 393
Colombia	2.6	1.6	1.2	0.1	25 205	34 752
Costa Rica	4.6	1.2	2.6	–0.8	34 679	39 649
Cuba	–4.3	4.4	–5.0	4.0	16 250	31 670
Dominica	2.5	–0.7	2.5	–1.4	13 092	13 248
Dominican Republic	–0.2	0.2	–1.6	–1.5	38 264	47 222
Ecuador	5.4	2.4	2.9	1.1	19 328	22 040
El Salvador	0.2	0.9	–0.1	0.0	18 786	24 452
Falkland Islands (Malvinas)	1.2	–1.1	0.2	–4.3	...	...
French Guiana	10.7	0.5	5.6	–2.6	34 147	26 510
Grenada	–1.2	–1.1	–0.7	–0.6	10 008	10 000
Guadeloupe	–0.4	2.4	–1.9	1.5	0	0
Guatemala	2.8	2.2	1.3	–0.1	18 489	17 351
Guyana	2.3	3.0	2.5	2.6	35 206	37 933
Haiti	–1.7	0.7	–3.7	–0.5	9 479	8 685
Honduras	3.4	1.4	0.2	–1.3	13 227	13 996
Jamaica	1.9	0.1	1.1	–0.8	14 907	11 670
Martinique	–2.0	3.8	–2.9	3.1	...	...
Mexico	2.0	2.4	0.2	0.9	26 221	28 246
Montserrat	...	...	...	...	18 750	18 750
Nicaragua	–1.5	5.7	–1.4	3.0	17 335	17 923
Panama	1.1	0.8	–1.0	–1.1	19 014	24 471
Paraguay	3.3	3.2	0.8	1.3	19 082	20 258
Peru	2.7	3.3	0.9	1.6	26 974	30 694
Puerto Rico	0.1	–0.6	–0.8	–1.3	14 043	17 308
Saint Kitts and Nevis	–2.6	0.2	–2.6	0.5	...	...
Saint Lucia	3.0	–2.6	1.5	–3.3	0	0
Saint Vincent and the Grenadines	–1.0	0.3	–1.8	–0.4	33 333	30 717
Suriname	–1.3	–1.1	–2.0	–1.9	38 159	38 455
Trinidad and Tobago	1.4	2.8	0.6	2.4	34 960	26 877
Uruguay	2.8	1.8	2.3	1.6	27 277	37 773
Venezuela, Bolivarian Republic of	2.2	2.5	–0.3	0.6	28 170	32 416
NEAR EAST AND NORTH AFRICA	**3.4**	**2.6**	**0.9**	**0.6**	**19 647**	**23 609**
Algeria	3.8	4.9	1.3	3.3	8 116	13 228
Bahrain	1.7	2.7	–1.8	0.0	...	...
Cyprus	0.5	2.5	–0.8	1.5	27 692	24 437
Egypt	3.7	4.1	2.0	2.2	59 184	71 912
Iran, Islamic Rep. of	5.3	3.0	2.6	1.7	16 903	23 871
Jordan	8.3	2.3	3.8	–0.9	14 621	10 731
Kuwait	17.9	10.3	18.8	6.1	57 223	21 361
Lebanon	6.0	–0.5	4.6	–2.5	21 075	24 864

TABLE A3 (cont.)

	Crop and livestock production		Per capita food production		Cereal yields	
	(Average annual rate of growth [%])				(hg/ha)	
	1985–1994	1995–2004	1985–1994	1995–2004	1992–1994	2002–2004
Libyan Arab Jamahiriya	2.1	2.1	–0.4	0.2	7 045	6 256
Morocco	6.9	3.7	4.9	2.1	9 110	11 921
Oman	2.2	3.1	–1.5	0.1	21 680	23 180
Qatar	12.4	7.3	7.6	5.3	31 212	41 304
Saudi Arabia	7.7	1.6	3.3	–1.4	44 002	37 611
Syrian Arab Republic	4.3	4.3	1.3	2.0	13 965	19 109
Tunisia	4.1	6.7	2.0	5.6	12 082	14 218
Turkey	1.9	1.7	0.0	0.1	20 966	22 982
United Arab Emirates	10.2	9.3	4.9	6.7	16 765	34 230
Yemen	4.2	3.1	0.0	–0.7	11 037	8 715
SUB-SAHARAN AFRICA	**3.6**	**2.4**	**0.8**	**–0.1**	**10 054**	**10 709**
Angola	2.7	4.0	0.2	1.1	3 212	5 023
Benin	5.4	6.5	0.9	4.3	9 298	10 604
Botswana	–0.1	–0.1	–3.0	–1.8	2 479	2 116
Burkina Faso	6.3	6.4	3.3	2.6	8 652	9 877
Burundi	1.7	0.6	–0.8	–0.7	13 484	13 333
Cameroon	3.9	2.9	1.4	0.6	10 005	17 098
Cape Verde	6.9	4.8	4.7	2.6	3 038	1 828
Central African Republic	2.7	3.0	0.9	1.5	9 349	10 471
Chad	5.8	3.7	2.7	0.9	6 591	7 125
Comoros	3.6	1.5	0.7	–1.4	13 194	13 341
Congo	1.3	2.3	–1.9	–0.7	7 519	7 796
Congo, Democratic Republic of the	2.4	–2.4	–0.7	–4.6	7 826	7 804
Côte d'Ivoire	3.0	1.8	–0.9	–0.1	9 209	11 382
Djibouti	2.9	1.8	–1.5	–0.6	15 833	16 250
Equatorial Guinea	3.0	–0.1	2.0	–2.2	...	...
Eritrea	35.9	0.3	35.7	–2.7	4 869	2 976
Ethiopia	1.2	3.9	–2.9	1.5	11 062	13 044
Gabon	1.9	1.6	–1.3	–0.9	18 048	16 410
Gambia	0.4	3.2	–3.3	0.3	11 969	11 071
Ghana	5.2	5.8	2.1	3.5	12 366	14 063
Guinea	3.2	2.7	–0.2	0.8	11 334	14 056
Guinea-Bissau	2.2	2.9	–0.4	0.0	14 227	11 376
Kenya	4.9	2.0	1.7	0.0	16 446	14 660
Lesotho	1.8	0.2	0.0	–0.1	8 014	9 628
Liberia	–4.9	6.1	–2.7	–2.0	10 370	9 167
Madagascar	1.0	1.0	–1.7	–1.7	19 278	20 594
Malawi	1.0	6.1	–2.8	6.1	9 559	11 353
Mali	4.9	3.3	1.7	–0.8	7 728	8 223
Mauritania	2.0	2.6	–0.3	–0.3	7 929	9 587
Mauritius	0.9	1.5	0.2	0.9	41 355	48 544
Mozambique	–0.1	4.9	–1.7	2.4	4 204	8 619
Namibia	3.7	–0.7	0.0	–3.0	2 769	4 105
Niger	7.1	4.7	3.7	1.2	3 130	4 151
Nigeria	7.7	2.5	4.5	–0.3	11 498	10 582
Réunion	3.3	1.3	1.5	–0.3	60 044	67 244

TABLE A3 (cont.)

	Crop and livestock production		Per capita food production		Cereal yields	
	(Average annual rate of growth [%])				(hg/ha)	
	1985–1994	1995–2004	1985–1994	1995–2004	1992–1994	2002–2004
Rwanda	−2.4	7.6	−1.5	2.4	11 496	10 011
Sao Tome and Principe	3.3	2.8	1.0	0.2	22 359	25 000
Senegal	5.7	2.2	3.3	−0.3	7 916	9 443
Seychelles	1.2	1.7	−0.3	0.7	...	...
Sierra Leone	1.0	−0.8	−0.8	−2.9	11 943	12 101
South Africa	2.5	1.5	0.6	0.4	19 013	26 756
Sudan	4.5	3.2	3.4	0.9	5 544	5 925
Swaziland	0.5	−0.4	−2.4	−1.9	14 072	11 138
Tanzania, United Republic of	0.9	2.2	−2.3	−0.4	11 617	14 756
Togo	4.9	2.8	0.7	−0.5	8 209	10 037
Uganda	3.1	2.8	−0.4	−0.3	15 220	16 509
Zambia	4.7	2.0	1.7	−0.2	14 945	15 136
Zimbabwe	3.9	1.2	2.7	−0.1	11 117	6 052
DEVELOPED MARKET ECONOMIES	0.8	1.1	0.1	0.4	43 703	49 313
Australia	1.0	3.0	−0.6	2.4	17 144	17 088
Austria	0.5	0.6	−0.1	0.5	52 195	56 120
Belgium-Luxembourg	...	...	...	...	65 634	0
Belgium	...	−0.4	...	−0.6	0	85 038
Canada	2.2	1.8	0.9	1.0	25 676	26 833
Denmark	0.3	0.4	0.1	0.1	51 739	60 026
Finland	−0.9	0.6	−1.3	0.4	33 568	32 309
France	−0.5	0.8	−1.0	0.4	65 146	70 341
Germany	−1.3	1.5	−1.7	1.4	56 246	63 240
Greece	1.7	−0.6	0.7	−1.0	36 738	35 387
Iceland	−1.4	0.8	−2.4	−0.1	...	...
Ireland	0.6	0.6	0.4	−0.5	62 375	70 298
Israel	−0.3	2.6	−2.1	0.4	27 229	31 087
Italy	0.2	−0.1	0.1	−0.1	47 447	48 864
Japan	−0.1	−1.1	−0.3	−1.3	55 850	59 489
Luxembourg	...	−3.2	...	−4.5	0	56 900
Malta	3.4	0.7	2.4	0.2	27 703	40 798
Netherlands	0.8	−1.0	0.2	−1.5	75 407	79 738
New Zealand	1.5	2.5	0.9	1.8	55 381	64 866
Norway	−0.5	−0.9	−1.0	−1.4	34 957	39 008
Portugal	2.7	1.3	2.8	1.1	20 704	27 746
Spain	0.1	3.1	−0.3	2.8	23 423	33 945
Sweden	−1.6	0.5	−2.1	0.4	40 032	48 817
Switzerland	−0.5	−0.1	−1.3	−0.2	61 368	60 052
United Kingdom	−0.3	−0.6	−0.6	−1.0	64 348	70 822
United States of America	2.2	1.3	1.1	0.2	50 746	61 384
COUNTRIES IN TRANSITION	−2.5	0.6	−2.9	0.9	19 636	23 175
Albania	1.3	1.9	0.9	2.1	24 652	31 433
Armenia	1.1	0.6	3.2	1.6	16 422	19 756
Azerbaijan	−13.9	3.2	−14.3	3.6	17 882	25 874

TABLE A3 (cont.)

	Crop and livestock production		Per capita food production		Cereal yields	
	(Average annual rate of growth [%])				(hg/ha)	
	1985–1994	1995–2004	1985–1994	1995–2004	1992–1994	2002–2004
Belarus	−10.6	0.9	−10.5	1.4	26 020	26 303
Bosnia and Herzegovina	−11.5	1.3	−6.2	−0.5	35 595	32 202
Bulgaria	−4.8	−0.3	−3.7	0.3	27 561	30 261
Croatia	−3.7	2.1	−1.3	2.4	41 243	44 320
Czech Republic	−20.4	0.8	−20.6	0.9	40 992	42 970
Estonia	−11.8	−2.3	−9.8	−1.1	16 874	21 841
Georgia	−0.2	−0.5	−0.1	0.7	19 271	20 487
Hungary	−2.8	2.2	−2.4	2.7	35 667	42 499
Kazakhstan	−14.7	−0.3	−14.6	0.5	10 555	10 567
Kyrgyzstan	−5.2	1.7	−5.0	0.8	23 492	27 670
Latvia	−20.1	−1.3	−18.4	−0.2	17 519	22 905
Lithuania	−17.4	0.8	−16.3	1.3	19 338	27 884
Macedonia, The former Yugoslav Republic of	−5.9	0.5	−5.4	−0.2	24 529	27 472
Moldova, Republic of	−6.5	−1.1	−6.7	−0.4	29 807	26 591
Poland	−1.8	1.7	−2.2	1.7	25 727	31 306
Romania	−1.9	2.2	−1.8	2.5	24 413	29 581
Russian Federation	−10.0	0.8	−9.8	1.3	16 122	18 907
Serbia and Montenegro	−0.6	0.8	−1.5	0.8	30 989	35 047
Slovakia	−3.1	0.3	−3.6	0.4	40 665	39 538
Slovenia	11.4	1.1	10.6	1.1	41 499	49 271
Tajikistan	−4.2	0.4	−6.4	−0.7	10 116	19 773
Turkmenistan	9.3	0.8	16.4	0.8	23 846	27 896
Ukraine	−10.7	0.5	−10.4	1.3	29 516	25 422
Uzbekistan	0.9	0.2	−0.2	−0.6	16 776	34 554

TABLE A4
Population and labour force indicators (2004)

	Total population	Rural population		Agricultural population		Economically active population	Economically active population in agriculture	
	(Thousands)	(Thousands)	(% of total)	(Thousands)	(% of total)	(Thousands)	(Thousands)	(%)
WORLD	6 373 555	3 270 558	51	2 599 791	41	3 125 649	1 347 123	43
DEVELOPED COUNTRIES	1 287 488	348 384	27	82 592	6	647 745	41 351	6
DEVELOPING COUNTRIES	5 086 067	2 922 174	57	2 517 199	49	2 477 904	1 305 772	53
ASIA AND THE PACIFIC	3 389 506	2 163 046	64	1 872 666	55	1 751 025	1 018 363	58
American Samoa	63	6	10	20	32	25	8	32
Bangladesh	149 664	112 836	75	77 454	52	76 756	39 723	52
Bhutan	2 325	2 121	91	2 176	94	1 127	1 055	94
British Virgin Islands	21	8	38	5	24	10	2	20
Brunei Darussalam	366	85	23	2	1	175	1	1
Cambodia	14 482	11 694	81	9 922	69	7 300	5 001	69
China, Hong Kong SAR	7 115	0	0	23	0	3 816	12	0
China, Macao SAR	468	5	1	0	0	250	0	0
China, Mainland	1 290 669	793 502	61	846 304	66	778 326	509 288	65
China, Taiwan Province of	22 640	1 127	5	3 090	14	10 219	710	7
Cook Islands	18	5	28	6	33	7	2	29
Fiji	847	401	47	322	38	354	134	38
French Polynesia	248	119	48	78	31	109	34	31
Guam	165	10	6	46	28	80	21	26
India	1 081 229	772 785	71	559 656	52	478 801	276 687	58
Indonesia	222 611	118 394	53	92 276	41	110 673	50 531	46
Kiribati	89	46	52	23	26	39	10	26
Korea, Democratic People's Rep. of	22 776	8 793	39	6 206	27	11 751	3 202	27
Korea, Republic of	47 951	9 440	20	3 255	7	25 169	1 944	8
Lao People's Dem. Rep.	5 787	4 565	79	4 385	76	2 933	2 223	76
Malaysia	24 876	8 724	35	3 739	15	10 935	1 740	16
Maldives	328	232	71	77	23	141	27	19
Marshall Islands	54	18	33	14	26	24	6	25
Micronesia, Federated States of	110	78	71	28	25	47	12	26
Mongolia	2 630	1 146	44	567	22	1 405	303	22
Myanmar	50 101	35 076	70	34 543	69	27 408	18 897	69
Nauru	13	0	0	3	23	6	1	17
Nepal	25 725	21 733	84	23 872	93	12 306	11 419	93
New Caledonia	233	90	39	79	34	124	42	34
Niue	2	1	50	1	50	1	0	0
Pakistan	157 315	103 181	66	76 917	49	59 145	26 682	45
Palau	21	7	33	5	24	9	2	22
Papua New Guinea	5 836	5 063	87	4 387	75	2 803	2 019	72
Philippines	81 408	31 091	38	30 078	37	34 860	12 942	37
Samoa	180	140	78	56	31	65	20	31
Singapore	4 315	0	0	5	0	2 149	2	0
Solomon Islands	491	408	83	352	72	253	181	72
Sri Lanka	19 218	15 178	79	8 668	45	8 910	3 948	44

TABLE A4 (cont.)

	Total population	Rural population		Agricultural population		Economically active population	Economically active population in agriculture	
	(Thousands)	(Thousands)	(% of total)	(Thousands)	(% of total)	(Thousands)	(Thousands)	(%)
Thailand	63 465	43 080	68	29 060	46	37 873	20 185	53
Timor-Leste	820	760	93	666	81	447	363	81
Tokelau	2	2	100	0	0	1	0	0
Tonga	105	70	67	33	31	39	12	31
Tuvalu	11	5	45	3	27	4	1	25
Vanuatu	217	167	77	74	34	97	33	34
Viet Nam	82 481	60 839	74	54 185	66	44 047	28 936	66
Wallis and Futuna Islands	15	15	100	5	33	6	2	33
LATIN AMERICA AND THE CARIBBEAN	**550 861**	**125 738**	**23**	**103 986**	**19**	**240 473**	**43 058**	**18**
Anguilla	12	0	0	3	25	6	1	17
Antigua and Barbuda	73	45	62	16	22	34	7	21
Argentina	38 871	3 755	10	3 585	9	16 381	1 455	9
Aruba	101	55	54	22	22	47	10	21
Bahamas	317	32	10	10	3	165	5	3
Barbados	271	129	48	10	4	152	5	3
Belize	261	135	52	77	30	94	28	30
Bermuda	82	0	0	2	2	42	1	2
Bolivia	8 973	3 244	36	3 762	42	3 755	1 619	43
Brazil	180 654	29 643	16	25 869	14	83 594	12 134	15
Cayman Islands	42	0	0	9	21	19	4	21
Chile	15 996	2 023	13	2 359	15	6 755	989	15
Colombia	44 914	10 359	23	8 386	19	20 020	3 666	18
Costa Rica	4 250	1 646	39	803	19	1 799	327	18
Cuba	11 328	2 756	24	1 679	15	5 688	727	13
Dominica	79	21	27	17	22	36	8	22
Dominican Republic	8 872	3 571	40	1 337	15	3 956	561	14
Ecuador	13 192	4 983	38	3 270	25	5 347	1 242	23
El Salvador	6 614	2 629	40	1 999	30	2 953	782	26
French Guiana	182	45	25	30	16	78	13	17
Grenada	80	47	59	18	23	37	8	22
Guadeloupe	443	3	1	11	2	206	5	2
Guatemala	12 661	6 740	53	6 006	47	4 792	2 089	44
Guyana	767	475	62	125	16	332	54	16
Haiti	8 437	5 226	62	5 070	60	3 710	2 232	60
Honduras	7 099	3 832	54	2 204	31	2 798	789	28
Jamaica	2 676	1 280	48	512	19	1 364	261	19
Martinique	395	17	4	13	3	188	6	3
Mexico	104 931	25 503	24	22 164	21	44 096	8 453	19
Montserrat	4	3	75	1	25	2	0	0
Netherlands Antilles	223	67	30	1	0	101	0	0
Nicaragua	5 597	2 363	42	1 003	18	2 285	392	17
Panama	3 177	1 353	43	665	21	1 353	248	18
Paraguay	6 018	2 539	42	2 314	38	2 323	756	33

TABLE A4 (cont.)

	Total population	Rural population		Agricultural population		Economically active population	Economically active population in agriculture	
	(Thousands)	(Thousands)	(% of total)	(Thousands)	(% of total)	(Thousands)	(Thousands)	(%)
Peru	27 567	7 098	26	7 767	28	10 818	3 074	28
Puerto Rico	3 898	81	2	89	2	1 476	26	2
Saint Kitts and Nevis	42	28	67	9	21	19	4	21
Saint Lucia	150	104	69	33	22	69	15	22
Saint Vincent and the Grenadines	121	49	40	27	22	54	12	22
Suriname	439	103	23	80	18	172	31	18
Trinidad and Tobago	1 307	315	24	103	8	607	48	8
Turks and Caicos Islands	21	11	52	5	24	10	2	20
United States Virgin Islands	112	7	6	24	21	52	11	21
Uruguay	3 439	248	7	368	11	1 564	189	12
Venezuela, Bolivarian Republic of	26 170	3 175	12	2 129	8	11 123	769	7
NEAR EAST AND NORTH AFRICA	**429 223**	**178 072**	**41**	**119 577**	**28**	**167 493**	**51 477**	**31**
Afghanistan	24 926	19 010	76	16 355	66	10 142	6 655	66
Algeria	32 339	13 160	41	7 406	23	12 033	2 800	23
Bahrain	739	71	10	6	1	352	3	1
Cyprus	808	248	31	58	7	403	29	7
Egypt	73 390	42 488	58	24 954	34	27 902	8 594	31
Iran, Islamic Republic of	69 788	22 785	33	17 157	25	26 727	6 602	25
Iraq	25 856	8 500	33	2 152	8	7 318	609	8
Jordan	5 614	1 158	21	567	10	1 933	195	10
Kuwait	2 595	103	4	27	1	1 391	15	1
Lebanon	3 708	439	12	105	3	1 412	40	3
Libyan Arab Jamahiriya	5 659	756	13	263	5	2 020	94	5
Morocco	31 064	13 026	42	10 408	34	12 979	4 296	33
Oman	2 935	648	22	983	33	1 082	362	33
Qatar	619	49	8	6	1	341	3	1
Saudi Arabia	24 919	3 030	12	1 844	7	8 554	633	7
Syrian Arab Republic	18 223	9 078	50	4 771	26	6 250	1 636	26
Tunisia	9 937	3 586	36	2 299	23	4 211	974	23
Turkey	72 320	24 133	33	20 484	28	34 269	14 854	43
United Arab Emirates	3 051	449	15	122	4	1 667	67	4
Yemen	20 733	15 355	74	9 610	46	6 507	3 016	46
SUB-SAHARAN AFRICA	**716 477**	**455 318**	**64**	**420 970**	**59**	**318 913**	**192 874**	**60**
Angola	14 078	8 956	64	9 962	71	6 390	4 521	71
Benin	6 918	3 782	55	3 463	50	3 163	1 583	50
Botswana	1 795	867	48	783	44	808	352	44
Burkina Faso	13 393	10 962	82	12 345	92	6 235	5 747	92
Burundi	7 068	6 349	90	6 341	90	3 739	3 355	90
Cameroon	16 296	7 789	48	7 807	48	6 807	3 728	55
Cape Verde	473	205	43	96	20	196	40	20
Central African Republic	3 912	2 213	57	2 705	69	1 827	1 264	69
Chad	8 854	6 612	75	6 319	71	4 021	2 870	71

TABLE A4 (cont.)

	Total population	Rural population		Agricultural population		Economically active population	Economically active population in agriculture	
	(Thousands)	(Thousands)	(% of total)	(Thousands)	(% of total)	(Thousands)	(Thousands)	(%)
Comoros	790	509	64	568	72	376	270	72
Congo	3 818	1 749	46	1 425	37	1 544	576	37
Congo, Democratic Republic of the	54 417	36 988	68	33 355	61	22 644	13 880	61
Côte d'Ivoire	16 897	9 243	55	7 571	45	6 934	3 107	45
Djibouti	712	114	16	547	77	354	272	77
Equatorial Guinea	507	258	51	348	69	209	143	68
Eritrea	4 297	3 426	80	3 278	76	2 101	1 603	76
Ethiopia	72 420	60 926	84	58 408	81	31 683	25 553	81
Gabon	1 351	205	15	444	33	611	201	33
Gambia	1 462	1 080	74	1 137	78	743	577	78
Ghana	21 377	11 550	54	11 801	55	10 773	6 021	56
Guinea	8 620	5 523	64	7 095	82	4 248	3 497	82
Guinea-Bissau	1 538	1 003	65	1 257	82	660	540	82
Kenya	32 420	19 257	59	23 873	74	17 070	12 570	74
Lesotho	1 800	1 474	82	691	38	721	277	38
Liberia	3 487	1 824	52	2 284	66	1 318	863	65
Madagascar	17 901	13 119	73	12 974	72	8 582	6 220	72
Malawi	12 337	10 283	83	9 327	76	5 876	4 777	81
Mali	13 409	8 989	67	10 549	79	6 253	4 920	79
Mauritania	2 980	1 105	37	1 546	52	1 329	689	52
Mauritius	1 233	694	56	124	10	546	56	10
Mozambique	19 182	12 088	63	14 538	76	10 041	8 065	80
Namibia	2 011	1 348	67	921	46	801	306	38
Niger	12 415	9 597	77	10 782	87	5 675	4 928	87
Nigeria	127 117	66 717	52	37 827	30	50 940	15 159	30
Réunion	767	64	8	19	2	323	8	2
Rwanda	8 481	6 781	80	7 644	90	4 512	4 067	90
Sao Tome and Principe	165	102	62	102	62	76	47	62
Senegal	10 339	5 136	50	7 488	72	4 652	3 369	72
Seychelles	82	41	50	63	77	39	30	77
Sierra Leone	5 168	3 166	61	3 103	60	1 920	1 153	60
Somalia	10 312	6 681	65	7 150	69	4 368	3 028	69
South Africa	45 214	19 153	42	5 621	12	18 897	1 570	8
Saint Helena	5	3	60	3	60	2	1	50
Sudan	34 333	20 654	60	19 708	57	13 806	7 925	57
Swaziland	1 083	827	76	343	32	376	119	32
Tanzania, United Republic of	37 671	23 907	63	28 729	76	19 337	15 214	79
Togo	5 017	3 218	64	2 873	57	2 142	1 227	57
Uganda	26 699	23 414	88	20 533	77	12 743	9 953	78
Zambia	10 924	7 008	64	7 313	67	4 597	3 078	67
Zimbabwe	12 932	8 359	65	7 787	60	5 905	3 555	60
DEVELOPED MARKET ECONOMIES	**880 421**	**196 057**	**22**	**26 396**	**3**	**436 566**	**12 761**	**3**
Andorra	73	7	10	6	8	33	3	9
Australia	19 913	1 484	7	853	4	10 174	436	4

TABLE A4 (cont.)

	Total population	Rural population		Agricultural population		Economically active population	Economically active population in agriculture	
	(Thousands)	*(Thousands)*	*(% of total)*	*(Thousands)*	*(% of total)*	*(Thousands)*	*(Thousands)*	*(%)*
Austria	8 120	2 778	34	352	4	3 745	162	4
Belgium	10 340	287	3	164	2	4 209	67	2
Canada	31 744	6 098	19	710	2	17 126	353	2
Denmark	5 375	781	15	174	3	2 891	93	3
Faeroe Islands	47	29	62	1	2	24	1	4
Finland	5 215	2 043	39	262	5	2 553	118	5
France	60 434	14 248	24	1 659	3	27 136	745	3
Germany	82 526	9 712	12	1 724	2	40 242	841	2
Gibraltar	27	0	0	2	7	12	1	8
Greece	10 977	4 243	39	1 285	12	4 827	707	15
Greenland	57	10	18	1	2	29	1	3
Iceland	292	20	7	22	8	166	12	7
Ireland	3 999	1 587	40	354	9	1 730	153	9
Israel	6 560	526	8	150	2	2 879	66	2
Italy	57 346	18 614	32	2 505	4	25 165	1 099	4
Japan	127 800	44 129	35	3 895	3	68 111	2 172	3
Liechtenstein	34	26	76	1	3	16	0	0
Luxembourg	459	36	8	8	2	197	3	2
Malta	396	32	8	5	1	151	2	1
Monaco	35	0	0	1	3	16	0	0
Netherlands	16 227	5 458	34	485	3	7 397	221	3
New Zealand	3 904	545	14	325	8	1 952	167	9
Norway	4 552	940	21	205	5	2 348	95	4
Portugal	10 072	4 551	45	1 262	13	5 121	570	11
San Marino	28	3	11	2	7	13	1	8
Spain	41 128	9 627	23	2 472	6	18 405	1 113	6
Sweden	8 886	1 481	17	275	3	4 772	131	3
Switzerland	7 164	2 350	33	422	6	3 795	143	4
United Kingdom	59 648	6 565	11	986	2	29 856	494	2
United States of America	297 043	57 847	19	5 828	2	151 475	2 791	2
COUNTRIES IN TRANSITION	**407 067**	**152 327**	**37**	**56 196**	**14**	**211 179**	**28 590**	**14**
Albania	3 194	1 790	56	1 457	46	1 633	745	46
Armenia	3 052	1 116	37	348	11	1 645	188	11
Azerbaijan	8 447	4 237	50	2 118	25	3 905	979	25
Belarus	9 852	2 851	29	1 113	11	5 364	606	11
Bosnia and Herzegovina	4 186	2 307	55	156	4	1 972	73	4
Bulgaria	7 829	2 338	30	458	6	4 067	222	5
Croatia	4 416	1 810	41	287	6	2 065	134	6
Czech Republic	10 226	2 630	26	742	7	5 697	413	7
Estonia	1 308	407	31	134	10	720	74	10
Georgia	5 074	2 452	48	905	18	2 626	468	18
Hungary	9 831	3 403	35	1 028	10	4 702	437	9
Kazakhstan	15 403	6 901	45	2 773	18	7 749	1 246	16
Kyrgyzstan	5 208	3 455	66	1 220	23	2 388	559	23
Latvia	2 286	794	35	245	11	1 264	135	11

TABLE A4 (cont.)

	Total population	Rural population		Agricultural population		Economically active population	Economically active population in agriculture	
	(Thousands)	(Thousands)	(% of total)	(Thousands)	(% of total)	(Thousands)	(Thousands)	(%)
Lithuania	3 422	1 153	34	430	13	1 766	183	10
Macedonia, The former Yugoslav Republic of	2 066	838	41	213	10	960	99	10
Moldova, Republic of	4 263	2 310	54	835	20	2 234	438	20
Poland	38 551	14 677	38	6 609	17	20 279	3 988	20
Romania	22 280	10 169	46	2 534	11	10 747	1 338	12
Russian Federation	142 397	38 250	27	13 453	9	78 053	7 374	9
Serbia and Montenegro	10 519	5 045	48	1 768	17	5 102	857	17
Slovakia	5 407	2 299	43	438	8	3 004	244	8
Slovenia	1 982	975	49	25	1	1 009	13	1
Tajikistan	6 298	4 770	76	1 961	31	2 671	832	31
Turkmenistan	4 940	2 688	54	1 572	32	2 289	728	32
Ukraine	48 151	15 845	33	6 748	14	25 162	3 188	13
Uzbekistan	26 479	16 817	64	6 626	25	12 106	3 029	25

TABLE 5
Land use

	Total land area	Forest and wood area	Agricultural area	Agricultural area per capita	Arable land	Permanent crops	Permanent pasture	Irrigated area	Fertilizer consumption
	(Thousand ha)			(ha/person)	(% of agricultural area)			(% of arable + permanent crops area)	(kg/ha arable land)
	2002	2000	2002	2002	2002	2002	2002	2002	2002
WORLD	13 039 650	3 868 796	5 006 880	0.80	28.0	2.7	69.2	18.0	100.8
DEVELOPED COUNTRIES	5 382 402	1 720 221	1 727 007	1.34	34.5	1.7	63.8	10.6	82.6
DEVELOPING COUNTRIES	7 657 248	2 148 575	3 279 873	0.66	24.6	3.3	72.1	23.0	114.3
ASIA AND THE PACIFIC	2 014 361	511 796	1 029 953	0.31	39.7	5.3	55.0	33.7	171.6
American Samoa	20	12	5	0.08	40.0	60.0	0.0	0.0	0.0
Bangladesh	13 017	1 334	9 029	0.06	88.8	4.5	6.6	54.5	177.5
Bhutan	4 700	3 016	580	0.26	25.0	3.4	71.6	24.2	0.0
British Virgin Islands	15	3	9	0.43	33.3	11.1	55.6	0.0	0.0
Brunei Darussalam	527	442	19	0.05	47.4	21.1	31.6	7.7	0.0
Cambodia	17 652	9 335	5 307	0.38	69.7	2.0	28.3	7.1	0.0
China, Hong Kong SAR	99	...	7	0.00	71.4	14.3	14.3	33.3	0.0
China, Macao SAR	2	...	...	0.00	...	...	...	...	...
China, Mainland	929 100	163 480	553 100	0.43	25.7	2.0	72.3	35.5	276.1
China, Taiwan Province of	3 541	...	850	0.04	72.5	27.5	0.0	68.2	630.5
Cocos (Keeling) Islands	1	...	...	...	...	...	...	...	...
Cook Islands	23	22	6	0.33	66.7	33.3	0.0	0.0	0.0
Fiji	1 827	815	460	0.55	43.5	18.5	38.0	1.1	61.5
French Polynesia	366	105	45	0.19	6.7	48.9	44.4	4.0	434.7
Guam	55	21	22	0.14	22.7	40.9	36.4	0.0	0.0
India	297 319	64 113	181 177	0.17	89.3	4.6	6.1	33.6	99.6
Indonesia	181 157	104 986	44 877	0.21	45.7	29.4	24.9	14.3	146.0
Kiribati	73	28	39	0.45	5.1	94.9	0.0	0.0	0.0
Korea, Democratic People's Rep. of	12 041	8 210	2 750	0.12	90.9	7.3	1.8	54.1	106.5
Korea, Republic of	9 873	6 248	1 933	0.04	87.1	10.0	2.9	60.6	409.7
Lao People's Dem. Rep.	23 080	12 561	1 879	0.34	49.0	4.3	46.7	17.5	7.6
Malaysia	32 855	19 292	7 870	0.33	22.9	73.5	3.6	4.8	683.3
Maldives	30	1	13	0.04	30.8	61.5	7.7	0.0	0.0
Marshall Islands	18	...	14	0.27	21.4	50.0	28.6	0.0	0.0
Micronesia, Federated States of	70	15	47	0.44	8.5	68.1	23.4	0.0	0.0
Mongolia	156 650	10 645	130 500	51.00	0.9	0.0	99.1	7.0	3.7
Myanmar	65 755	34 419	10 925	0.22	90.3	6.9	2.9	18.8	13.4
Northern Mariana Islands	46	14	13	0.17	46.2	15.4	38.5	0.0	0.0

TABLE 5 (cont.)

	Total land area	Forest and wood area	Agricultural area	Agricultural area per capita	Arable land	Permanent crops	Permanent pasture	Irrigated area	Fertilizer consumption
	(Thousand ha)			(ha/person)	(% of agricultural area)			(% of arable + permanent crops area)	(kg/ha arable land)
	2002	2000	2002	2002	2002	2002	2002	2002	2002
Nauru	2	...	...	0.00	...	...	...	...	...
Nepal	14 300	3 900	5 031	0.20	63.6	1.9	34.5	34.5	27.8
New Caledonia	1 828	372	248	1.11	2.0	1.6	96.4	111.1	180.0
Niue	26	6	8	4.00	50.0	37.5	12.5	0.0	0.0
Norfolk Island	4	...	1	...	0.0	0.0	100.0	...	...
Pakistan	77 088	2 361	27 120	0.18	79.1	2.5	18.4	80.5	138.1
Palau	46	35	9	0.45	44.4	22.2	33.3	0.0	0.0
Papua New Guinea	45 286	30 601	1 045	0.19	21.1	62.2	16.7	0.0	53.6
Philippines	29 817	5 789	12 200	0.16	46.7	41.0	12.3	14.5	126.8
Samoa	283	105	131	0.74	45.8	52.7	1.5	0.0	58.3
Singapore	67	2	2	0.00	50.0	50.0	0.0	0.0	2 418.0
Solomon Islands	2 799	2 536	115	0.25	15.7	49.6	34.8	0.0	0.0
Sri Lanka	6 463	1 940	2 356	0.12	38.9	42.4	18.7	33.3	310.3
Thailand	51 089	14 762	20 167	0.32	78.7	17.4	4.0	25.6	107.2
Timor-Leste	1 487	...	287	0.39	24.4	23.3	52.3	0.0	0.0
Tokelau	1	...	...	0.00	...	...	...	...	...
Tonga	72	4	52	0.50	32.7	59.6	7.7	0.0	0.0
Tuvalu	3	...	...	0.00	...	...	...	...	...
Vanuatu	1 219	447	162	0.78	18.5	55.6	25.9	0.0	0.0
Viet Nam	32 549	9 819	9 537	0.12	70.3	23.0	6.7	33.7	294.8
Wallis and Futuna Islands	20	...	6	0.40	16.7	83.3	0.0	0.0	0.0
LATIN AMERICA AND THE CARIBBEAN	2 018 063	964 355	784 270	1.46	18.9	2.6	78.5	11.0	89.3
Antigua and Barbuda	44	9	14	0.19	57.1	14.3	28.6	0.0	0.0
Argentina	273 669	34 648	177 000	4.66	19.0	0.7	80.2	4.5	21.9
Aruba	19	...	2	0.02	100.0	0.0	0.0	0.0	0.0
Bahamas	1 001	842	14	0.05	57.1	28.6	14.3	8.3	100.0
Barbados	43	2	19	0.07	84.2	5.3	10.5	5.9	50.7
Belize	2 280	1 348	152	0.61	46.1	21.1	32.9	2.9	67.1
Bermuda	5	...	1	0.01	100.0	0.0	0.0	0.0	100.0
Bolivia	108 438	53 068	36 937	4.27	7.9	0.6	91.6	4.2	4.7
Brazil	845 942	543 905	263 580	1.50	22.4	2.9	74.7	4.4	130.2
Cayman Islands	26	13	3	0.08	33.3	0.0	66.7	0.0	0.0
Chile	74 880	15 536	15 242	0.98	13.0	2.1	84.9	82.4	229.6
Colombia	103 870	49 601	45 911	1.05	5.0	3.4	91.6	23.4	301.6
Costa Rica	5 106	1 968	2 865	0.70	7.9	10.5	81.7	20.6	673.6
Cuba	10 982	2 348	6 655	0.59	40.1	16.8	43.1	23.0	45.7
Dominica	75	46	22	0.28	22.7	68.2	9.1	0.0	108.6
Dominican Republic	4 838	1 376	3 696	0.43	29.7	13.5	56.8	17.2	81.8
Ecuador	27 684	10 557	8 075	0.63	20.1	16.9	63.0	29.0	141.7

TABLE 5 (cont.)

	Total land area	Forest and wood area	Agricultural area	Agricultural area per capita	Arable land	Permanent crops	Permanent pasture	Irrigated area	Fertilizer consumption
	(Thousand ha)			(ha/person)	(% of agricultural area)			(% of arable + permanent crops area)	(kg/ha arable land)
	2002	2000	2002	2002	2002	2002	2002	2002	2002
El Salvador	2 072	121	1 704	0.27	38.7	14.7	46.6	4.9	83.8
Falkland Islands (Malvinas)	1 217	...	1 130	376.67	0.0	0.0	100.0	...	...
French Guiana	8 815	7 926	23	0.13	52.2	17.4	30.4	12.5	100.0
Grenada	34	5	13	0.16	15.4	76.9	7.7	0.0	0.0
Guadeloupe	169	82	48	0.11	39.6	12.5	47.9	24.0	973.7
Guatemala	10 843	2 850	4 507	0.37	30.2	12.1	57.7	6.8	136.9
Guyana	19 685	16 879	1 740	2.28	27.6	1.7	70.7	29.4	37.2
Haiti	2 756	88	1 590	0.19	49.1	20.1	30.8	6.8	17.9
Honduras	11 189	5 383	2 936	0.43	36.4	12.3	51.4	5.6	47.0
Jamaica	1 083	325	513	0.20	33.9	21.4	44.6	8.8	128.7
Martinique	106	47	33	0.08	30.3	33.3	36.4	33.3	1 770.0
Mexico	190 869	55 205	107 300	1.05	23.1	2.3	74.6	23.2	69.0
Montserrat	10	3	3	1.00	66.7	0.0	33.3	0.0	0.0
Netherlands Antilles	80	1	8	0.04	100.0	0.0	0.0	0.0	0.0
Nicaragua	12 140	3 278	6 976	1.31	27.6	3.4	69.0	4.3	27.9
Panama	7 443	2 876	2 230	0.73	24.6	6.6	68.8	5.0	52.4
Paraguay	39 730	23 372	24 815	4.32	12.2	0.4	87.4	2.2	50.7
Peru	128 000	65 215	31 410	1.17	11.8	1.9	86.3	27.7	74.1
Puerto Rico	887	229	294	0.08	11.9	16.7	71.4	47.6	0.0
Saint Kitts and Nevis	36	4	10	0.24	70.0	10.0	20.0	0.0	242.9
Saint Lucia	61	9	20	0.14	20.0	70.0	10.0	16.7	335.8
Saint Vincent and the Grenadines	39	6	16	0.13	43.8	43.8	12.5	7.1	304.7
Suriname	15 600	14 113	88	0.20	64.8	11.4	23.9	76.1	98.2
Trinidad and Tobago	513	259	133	0.10	56.4	35.3	8.3	3.3	43.4
Turks and Caicos Islands	43	...	1	0.05	100.0	0.0	0.0	0.0	0.0
Uruguay	17 502	1 292	14 883	4.39	8.7	0.3	91.0	13.5	99.2
United States Virgin Islands	34	14	10	0.09	40.0	10.0	50.0	0.0	150.0
Venezuela, Bolivarian Republic of	88 205	49 506	21 648	0.86	12.0	3.7	84.3	16.9	115.5
NEAR EAST AND NORTH AFRICA	**1 262 615**	**28 820**	**458 642**	**1.11**	**19.3**	**2.5**	**78.2**	**28.7**	**73.1**
Afghanistan	65 209	1 351	38 054	1.66	20.8	0.4	78.8	29.6	2.6
Algeria	238 174	2 145	40 065	1.28	19.1	1.5	79.4	6.8	12.8
Bahrain	71	...	10	0.01	20.0	40.0	40.0	66.7	50.0
Cyprus	924	172	117	0.15	61.5	35.0	3.4	35.4	214.0
Egypt	99 545	72	3 400	0.05	85.3	14.7	0.0	100.0	437.5
Iran, Islamic Republic of	163 620	7 299	61 088	0.90	24.6	3.4	72.0	43.9	86.0

TABLE 5 (cont.)

	Total land area	Forest and wood area	Agricultural area	Agricultural area per capita	Arable land	Permanent crops	Permanent pasture	Irrigated area	Fertilizer consumption
	(Thousand ha)			(ha/person)	(% of agricultural area)			(% of arable + permanent crops area)	(kg/ha arable land)
	2002	2000	2002	2002	2002	2002	2002	2002	2002
Iraq	43 737	799	10 090	0.41	57.0	3.4	39.6	57.9	111.1
Jordan	8 893	86	1 142	0.21	25.8	9.2	65.0	18.8	113.6
Kuwait	1 782	5	151	0.06	8.6	1.3	90.1	86.7	80.8
Lebanon	1 023	36	329	0.09	51.7	43.5	4.9	33.2	231.9
Libyan Arab Jamahiriya	175 954	358	15 450	2.84	11.7	2.2	86.1	21.9	34.1
Morocco	44 630	3 025	30 283	1.01	27.7	2.9	69.3	14.5	47.5
Oman	30 950	1	1 081	0.39	3.5	4.0	92.5	76.5	321.9
Qatar	1 100	1	71	0.12	25.4	4.2	70.4	61.9	0.0
Saudi Arabia	214 969	1 504	173 794	7.39	2.1	0.1	97.8	42.7	105.9
Syrian Arab Republic	18 378	461	13 759	0.79	33.4	6.0	60.6	24.6	70.3
Tunisia	15 536	510	9 763	1.00	28.4	21.9	49.7	7.8	36.8
Turkey	76 963	10 225	41 690	0.59	62.2	6.2	31.6	18.3	67.2
United Arab Emirates	8 360	321	571	0.19	13.1	33.5	53.4	28.6	466.7
Yemen	52 797	449	17 734	0.92	8.7	0.7	90.6	30.0	7.5
SUB-SAHARAN AFRICA	2 362 209	643 604	1 007 008	1.47	16.0	2.1	81.9	3.7	14.6
Angola	124 670	69 756	57 300	4.35	5.2	0.5	94.2	2.3	0.0
Benin	11 062	2 650	3 365	0.51	75.8	7.9	16.3	0.4	18.8
Botswana	56 673	12 427	25 980	14.68	1.5	0.0	98.5	0.3	12.2
Burkina Faso	27 360	7 089	10 400	0.82	41.8	0.5	57.7	0.6	0.4
Burundi	2 568	94	2 170	0.33	45.4	16.8	45.6	5.5	2.6
Cameroon	46 540	23 858	9 160	0.58	65.1	13.1	21.8	0.5	5.9
Cape Verde	403	85	70	0.15	60.0	4.3	35.7	6.7	5.2
Central African Republic	62 298	22 907	5 149	1.35	37.5	1.8	60.7	0.0	0.3
Chad	125 920	12 692	48 630	5.83	7.4	0.1	92.5	0.6	4.9
Comoros	223	8	147	0.20	54.4	35.4	10.2	0.0	3.8
Congo	34 150	22 060	10 240	2.82	1.9	0.5	97.7	0.4	1.2
Congo, Democratic Republic of the	226 705	135 207	22 800	0.45	29.4	4.8	65.8	0.1	1.6
Côte d'Ivoire	31 800	7 117	19 900	1.22	15.6	19.1	65.3	1.1	35.2
Djibouti	2 318	6	1 301	1.88	0.1	0.0	99.9	100.0	0.0
Equatorial Guinea	2 805	1 752	334	0.69	38.9	29.9	31.1	0.0	0.0
Eritrea	10 100	1 585	7 470	1.87	6.7	0.0	93.3	4.2	7.4
Ethiopia	100 000	4 593	30 671	0.44	32.4	2.4	65.2	1.8	15.1
Gabon	25 767	21 826	5 160	3.95	6.3	3.3	90.4	3.0	0.9
Gambia	1 000	481	714	0.51	35.0	0.7	64.3	0.8	3.2
Ghana	22 754	6 335	14 681	0.72	28.5	14.6	56.9	0.2	7.4
Guinea	24 572	6 929	12 240	1.46	7.4	5.2	87.4	6.2	3.6
Guinea-Bissau	2 812	2 187	1 628	1.12	18.4	15.2	66.3	3.1	8.0
Kenya	56 914	17 096	26 462	0.84	17.4	2.1	80.5	1.7	31.0

TABLE 5 (cont.)

	Total land area	Forest and wood area	Agricultural area	Agricultural area per capita	Arable land	Permanent crops	Permanent pasture	Irrigated area	Fertilizer consumption
	(Thousand ha)			(ha/person)	(% of agricultural area)			(% of arable + permanent crops area)	(kg/ha arable land)
	2002	2000	2002	2002	2002	2002	2002	2002	2002
Lesotho	3 035	14	2 334	1.30	14.1	0.2	85.7	0.3	34.2
Liberia	9 632	3 481	2 600	0.80	14.6	8.5	76.9	0.5	0.0
Madagascar	58 154	11 727	27 550	1.63	10.7	2.2	87.1	30.7	3.1
Malawi	9 408	2 562	4 290	0.36	53.6	3.3	43.1	1.2	83.9
Mali	122 019	13 186	34 700	2.75	13.4	0.1	86.5	2.9	9.0
Mauritania	102 522	317	39 750	14.16	1.2	0.0	98.7	9.8	5.9
Mauritius	203	16	113	0.09	88.5	5.3	6.2	20.8	250.0
Mozambique	78 409	30 601	48 435	2.61	8.7	0.5	90.8	2.4	5.9
Namibia	82 329	8 040	38 820	19.80	2.1	0.0	97.9	0.9	0.4
Niger	126 670	1 328	16 500	1.43	27.2	0.1	72.7	1.5	1.1
Nigeria	91 077	13 517	72 200	0.60	41.8	3.9	54.3	0.7	5.5
Réunion	250	71	49	0.07	69.4	6.1	24.5	32.4	147.1
Rwanda	2 467	307	1 850	0.22	60.3	14.5	25.1	0.4	13.7
Saint Helena	31	2	12	2.40	33.3	0.0	66.7	0.0	0.0
Sao Tome and Principe	96	27	55	0.35	12.7	85.5	1.8	18.5	0.0
Senegal	19 253	6 205	8 156	0.83	30.2	0.6	69.3	2.8	13.6
Seychelles	45	30	7	0.09	14.3	85.7	0.0	0.0	17.0
Sierra Leone	7 162	1 055	2 800	0.59	19.1	2.3	78.6	5.0	0.6
Somalia	62 734	7 515	44 071	4.65	2.4	0.1	97.6	18.7	0.5
South Africa	121 447	8 917	99 640	2.23	14.8	1.0	84.2	9.5	65.4
Sudan	237 600	61 627	133 833	4.07	12.1	0.3	87.6	11.7	4.3
Swaziland	1 720	522	1 390	1.30	12.8	0.9	86.3	36.8	39.3
Tanzania, United Republic of	88 359	38 811	40 100	1.11	10.0	2.7	87.3	3.3	1.8
Togo	5 439	510	3 630	0.76	69.1	3.3	27.5	0.7	6.8
Uganda	19 710	4 190	12 312	0.49	41.4	17.1	41.5	0.1	1.8
Zambia	74 339	31 246	35 289	3.30	14.9	0.1	85.0	0.9	12.4
Zimbabwe	38 685	19 040	20 550	1.60	15.7	0.6	83.7	3.5	34.2
DEVELOPED MARKET ECONOMIES	3 070 643	783 052	1 095 472	1.24	32.1	2.0	65.9	11.3	118.2
Andorra	48	...	26	0.38	3.8	0.0	96.2	0.0	0.0
Australia	768 230	154 539	447 000	22.87	10.8	0.1	89.1	5.2	47.2
Austria	8 273	3 886	3 397	0.42	40.9	2.1	57.0	0.3	149.7
Belgium–Luxembourg	3 282	728	1 519	0.14	53.8	1.6	44.6	4.8	353.7
Canada	922 097	244 571	67 505	2.16	67.6	9.6	22.8	1.5	57.2
Denmark	4 243	455	2 666	0.50	85.4	0.3	14.3	19.6	130.5
Faeroe Islands	140	...	3	0.06	100.0	0.0	0.0	0.0	0.0
Finland	30 459	21 935	2 228	0.43	98.7	0.4	0.9	2.9	133.2
France	55 010	15 341	29 555	0.49	62.4	3.8	33.7	13.3	215.1
Germany	34 895	10 740	16 967	0.21	69.5	1.2	29.3	4.0	220.0
Gibraltar	1	...	...	0.00	...	...	...	...	...

TABLE 5 (cont.)

	Total land area	Forest and wood area	Agricultural area	Agricultural area per capita	Arable land	Permanent crops	Permanent pasture	Irrigated area	Fertilizer consumption
	(Thousand ha)			(ha/person)	(% of agricultural area)			(% of arable + permanent crops area)	(kg/ha arable land)
	2002	2000	2002	2002	2002	2002	2002	2002	2002
Greece	12 890	3 599	8 446	0.77	32.2	13.4	54.5	37.2	149.1
Greenland	41 045	...	235	4.20	0.0	0.0	100.0	...	...
Iceland	10 025	31	2 281	7.95	0.3	0.0	99.7	0.0	2 555.4
Ireland	6 889	659	4 408	1.13	25.4	0.0	74.5	0.0	523.6
Israel	2 171	132	566	0.09	59.7	15.2	25.1	45.8	240.5
Italy	29 411	10 003	15 443	0.27	53.7	18.0	28.4	24.9	172.9
Japan	36 450	24 081	5 190	0.04	85.1	6.6	8.2	54.7	290.6
Liechtenstein	16	7	9	0.27	44.4	0.0	55.6	0.0	0.0
Malta	32	...	10	0.03	90.0	10.0	0.0	20.0	77.8
Netherlands	3 388	375	1 949	0.12	47.0	1.7	51.3	59.5	366.8
New Zealand	26 799	7 946	17 235	4.48	8.7	10.9	80.4	8.5	568.6
Norway	30 625	8 868	1 033	0.23	84.3	0.0	15.7	14.6	211.3
Portugal	9 150	3 666	4 142	0.41	48.0	17.3	34.7	24.0	104.0
Saint Pierre and Miquelon	23	...	3	0.50	100.0	0.0	0.0	0.0	0.0
San Marino	6	...	1	0.04	100.0	0.0	0.0	0.0	390.0
Spain	49 944	14 370	30 195	0.74	45.5	16.5	38.0	20.2	157.2
Sweden	41 162	27 134	3 129	0.35	85.6	0.1	14.3	4.3	100.0
Switzerland	3 955	1 199	1 525	0.21	26.8	1.6	71.6	5.8	227.5
United Kingdom	24 088	2 794	16 943	0.29	34.0	0.3	65.7	2.9	313.1
United States of America	915 896	225 993	411 863	1.42	42.7	0.5	56.8	12.6	109.6
COUNTRIES IN TRANSITION	2 311 759	937 169	631 535	1.54	38.8	1.1	60.1	9.7	31.5
Albania	2 740	991	1 140	0.36	50.7	10.6	38.7	48.6	61.2
Armenia	2 820	351	1 395	0.45	35.5	4.7	59.9	50.0	22.8
Azerbaijan	8 260	1 094	4 692	0.57	38.0	4.8	57.2	72.4	9.9
Belarus	20 748	9 402	8 924	0.90	62.8	1.4	35.8	2.3	133.4
Bosnia and Herzegovina	5 120	2 273	2 123	0.51	47.0	4.5	48.5	0.3	32.7
Bulgaria	11 063	3 690	5 325	0.67	63.0	4.3	32.7	16.5	49.5
Croatia	5 592	1 783	3 143	0.71	46.5	4.0	49.5	0.3	117.6
Czech Republic	7 728	2 632	4 273	0.42	71.8	5.5	22.7	0.7	120.2
Estonia	4 239	2 060	698	0.52	87.8	2.6	9.6	0.6	44.1
Georgia	6 949	2 988	3 004	0.58	26.6	8.8	64.6	44.1	35.5
Hungary	9 210	1 840	5 867	0.59	78.6	3.2	18.1	4.8	108.7
Kazakhstan	269 970	12 148	206 769	13.37	10.4	0.1	89.5	10.8	3.0
Kyrgyzstan	19 180	1 003	10 776	2.13	12.5	0.6	86.9	76.0	20.5
Latvia	6 205	2 923	2 474	1.06	74.1	1.2	24.8	1.1	27.3
Lithuania	6 268	1 994	3 487	1.01	84.0	1.7	14.3	0.2	66.2
Macedonia, The former Yugoslav Republic of	2 543	906	1 242	0.61	45.6	3.7	50.7	9.0	39.4

TABLE 5 (cont.)

	Total land area	Forest and wood area	Agricultural area	Agricultural area per capita	Arable land	Permanent crops	Permanent pasture	Irrigated area	Fertilizer consumption
	(Thousand ha)			(ha/person)	(% of agricultural area)			(% of arable + permanent crops area)	(kg/ha arable land)
	2002	2000	2002	2002	2002	2002	2002	2002	2002
Moldova, Republic of	3 288	325	2 534	0.59	72.7	11.8	15.4	14.0	5.5
Poland	30 629	9 047	18 345	0.47	75.9	1.7	22.5	0.7	108.6
Romania	22 987	6 448	14 837	0.66	63.3	3.4	33.3	31.1	34.7
Russian Federation	1 688 850	851 392	216 651	1.50	57.0	0.8	42.2	3.7	11.9
Serbia and Montenegro	10 200	2 887	5 586	0.53	60.8	5.9	33.3	0.8	90.6
Slovakia	4 808	2 177	2 433	0.45	58.9	5.2	35.9	11.7	86.8
Slovenia	2 014	1 107	505	0.25	33.3	5.9	60.8	1.5	416.0
Tajikistan	13 996	400	4 255	0.69	21.9	3.0	75.2	68.0	30.0
Turkmenistan	46 993	3 755	32 615	6.80	5.7	0.2	94.1	94.0	52.9
Ukraine	57 935	9 584	41 396	0.85	78.6	2.2	19.2	6.8	18.1
Uzbekistan	41 424	1 969	27 046	1.05	16.6	1.3	82.2	88.7	160.2

TABLE 6
Trade indicators (average 2001–03)

	Agricultural exports (Million $)	Agricultural imports (Million $)	Agricultural exports as share of total exports (%)	Agricultural imports as share of total imports (%)	Net food imports (Thousand $)	Agricultural exports relative to agricultural GDP (%)
WORLD	459 493	482 258	6.9	7.1	15 658 754	36.2
DEVELOPED COUNTRIES	321 039	340 935	7.0	7.0	16 133 838	74.4
DEVELOPING COUNTRIES	138 356	141 324	6.6	7.4	–475 084	19.4
ASIA AND THE PACIFIC	57 506	66 497	4.3	5.2	779 589	12.8
American Samoa	0	16	0.1	4.2	12 981	...
Bangladesh	99	1 543	1.7	17.1	860 245	0.9
Bhutan	14	20	12.6	11.0	2 605	7.1
British Virgin Islands	0	13	1.5	9.5	5 435	...
Brunei Darussalam	2	214	0.1	14.4	161 961	...
Cambodia	26	293	1.6	14.4	105 356	1.9
China, Hong Kong SAR	3 607	8 089	1.8	3.8	3 393 527	2 981.7
China, Macao SAR	48	344	2.0	13.5	131 778	...
China, Mainland	13 824	12 851	4.1	4.0	–7 133 624	7.1
China, Taiwan Province of	958	5 802	0.7	5.0	2 545 330	...
Cook Islands	1	13	5.3	11.4	9 952	...
Fiji	174	132	29.7	14.0	–35 890	70.0
French Polynesia	15	227	8.4	17.3	183 417	...
Guam	0	48	0.2	9.8	33 834	...
India	5 753	4 282	11.2	7.0	–3 231 471	5.1
Indonesia	5 856	4 219	9.5	12.3	792 715	20.0
Kiribati	2	12	26.1	22.4	8 863	38.4
Korea, Democratic People's Rep. of	21	338	2.3	26.0	256 209	...
Korea, Republic of	1 726	8 963	1.0	5.7	4 672 519	8.9
Lao People's Dem. Rep.	66	86	18.7	17.5	35 076	7.1
Malaysia	7 492	4 221	7.8	5.4	1 361 848	86.2
Maldives	0	90	0.1	21.5	72 488	...
Marshall Islands	1	0	16.8	...	–18	13.7
Micronesia, Federated States of	4	13	10.9	15.4	9 760	...
Mongolia	97	119	17.8	17.0	67 834	39.2
Myanmar	377	305	13.4	12.5	–205 714	...
Nauru	...	2	...	8.5	1 351	...
Nepal	84	196	12.6	13.1	42 729	4.0
New Caledonia	3	155	0.4	8.1	117 975	...
Niue	0	1	90.0	11.0	255	...
Norfolk Island	1	3	32.2	11.1	1 329	...
Pakistan	1 081	1 618	10.5	14.1	–327 451	6.5
Papua New Guinea	286	188	15.2	17.9	13 973	38.1
Philippines	1 606	2 715	4.5	7.3	1 161 199	14.2
Samoa	5	27	34.0	20.5	22 752	13.0
Singapore	2 602	3 935	2.0	3.3	1 248 166	2 727.1
Solomon Islands	33	15	36.8	14.0	7 073	...
Sri Lanka	838	783	17.3	12.6	–88 787	28.3
Thailand	8 649	3 142	12.4	4.8	–4 606 031	70.4
Timor-Leste	10	45	26.9	30.7	30 817	9.7

TABLE 6 (cont.)

	Agricultural exports	Agricultural imports	Agricultural exports as share of total exports	Agricultural imports as share of total imports	Net food imports	Agricultural exports relative to agricultural GDP
	(Million $)	(Million $)	(%)	(%)	(Thousand $)	(%)
Tonga	10	22	36.0	25.6	9 296	23.2
Tuvalu	0	2	0.6	11.2	1 953	...
Vanuatu	10	15	66.9	16.5	7 419	28.8
Viet Nam	2 124	1 378	12.4	6.9	−983 050	26.3
Wallis and Futuna Islands	0	2	0.0	4.9	1 606	...
LATIN AMERICA AND THE CARIBBEAN	**56 409**	**30 220**	**19.7**	**9.9**	**−17 473 197**	**52.3**
Antigua and Barbuda	1	29	0.2	6.2	21 340	1.5
Argentina	11 959	816	43.7	5.7	−7 295 746	99.5
Aruba	78	235	4.1	10.5	125 133	...
Bahamas	127	297	2.5	13.4	217 319	...
Barbados	72	181	28.8	16.3	86 513	59.3
Belize	116	69	63.7	13.5	−62 397	96.6
Bermuda	0	84	0.0	2.0	61 091	...
Bolivia	447	242	32.3	14.2	−128 482	43.6
Brazil	17 900	3 349	28.0	6.4	−9 678 044	70.0
Cayman Islands	...	44	...	6.7	24 867	...
Chile	3 442	1 236	17.8	6.8	−1 552 225	66.7
Colombia	2 744	1 606	22.2	12.2	−816 933	28.0
Costa Rica	1 643	526	30.4	7.6	−1 022 667	123.4
Cuba	836	850	51.5	17.5	90 276	...
Dominica	16	28	39.5	22.7	5 382	42.1
Dominican Republic	585	708	66.1	12.2	149 999	25.8
Ecuador	1 739	563	33.2	9.2	−954 958	84.5
El Salvador	400	739	13.4	14.2	229 411	31.5
Falkland Islands (Malvinas)	7	...	94.7	...	42	...
Grenada	20	36	40.2	15.8	10 954	62.3
Guatemala	1 286	835	53.8	13.8	−412 024	25.0
Guyana	165	91	32.8	15.9	−81 072	83.5
Haiti	19	370	6.5	33.3	284 497	2.2
Honduras	638	507	48.1	16.6	−117 685	79.8
Jamaica	272	412	21.2	11.4	155 268	58.9
Mexico	8 077	11 518	9.4	10.2	2 351 960	34.6
Montserrat	0	4	1.3	16.9	2 628	...
Netherlands Antilles	10	177	0.6	5.9	107 341	...
Nicaragua	378	288	40.7	14.8	−108 658	52.7
Panama	279	415	31.9	13.8	98 562	31.1
Paraguay	752	237	54.3	9.6	−122 379	51.1
Peru	760	1 060	9.6	13.9	91 485	14.2
Saint Kitts and Nevis	10	27	15.7	15.1	11 032	111.5
Saint Lucia	32	94	74.9	30.4	57 215	90.8
Saint Vincent and the Grenadines	29	38	72.1	20.9	5 218	101.0
Suriname	48	90	9.1	15.5	14 687	54.5
Trinidad and Tobago	233	347	5.4	9.4	159 688	199.3

TABLE 6 (cont.)

	Agricultural exports (Million $)	Agricultural imports (Million $)	Agricultural exports as share of total exports (%)	Agricultural imports as share of total imports (%)	Net food imports (Thousand $)	Agricultural exports relative to agricultural GDP (%)
Uruguay	1 020	336	50.0	14.3	–552 715	82.5
Venezuela, Bolivarian Republic of	269	1 739	1.1	14.6	1 070 882	6.3
NEAR EAST AND NORTH AFRICA	**11 387**	**32 311**	**3.3**	**12.3**	**15 939 670**	**11.2**
Algeria	39	2 773	0.2	23.2	2 408 313	0.7
Bahrain	40	534	0.7	11.1	394 413	...
Cyprus	303	549	32.9	13.3	202 056	...
Egypt	778	3 151	10.6	20.6	2 169 704	5.6
Iran, Islamic Republic of	1 215	2 668	4.3	12.0	627 347	8.4
Jordan	390	898	14.3	17.4	467 795	218.6
Kuwait	39	991	0.2	11.0	838 333	...
Lebanon	196	1 230	17.3	17.7	811 943	10.2
Libyan Arab Jamahiriya	15	1 165	0.2	27.2	871 944	...
Morocco	835	1 692	10.5	13.9	487 396	13.6
Occupied Palestinian Territory	61	501	22.0	19.8	371660	...
Oman	577	1 230	5.1	20.0	489 848	...
Qatar	10	391	0.1	9.5	332 706	...
Saudi Arabia	404	5 207	0.5	16.0	4 237 948	4.2
Syrian Arab Republic	772	811	12.2	16.0	118 173	16.4
Tunisia	438	945	6.1	9.5	460 546	17.5
Turkey	4 134	3 224	11.0	6.0	–2 159 238	19.6
United Arab Emirates	1 054	3 458	1.3	6.3	2 117 964	...
Yemen	87	893	2.5	29.9	690 819	5.7
SUB-SAHARAN AFRICA	**13 055**	**12 296**	**13.7**	**13.3**	**278 855**	**20.2**
Angola	2	658	0.0	19.9	386 113	0.2
Benin	203	235	44.9	33.6	130 955	20.0
Botswana	77	237	3.2	14.2	140 640	52.2
Burkina Faso	162	139	69.0	18.6	69 712	15.1
Burundi	29	22	79.6	15.9	–7 915	9.5
Cameroon	497	262	24.3	12.8	–115 824	11.2
Cape Verde	0	83	1.9	31.9	59 463	0.3
Central African Republic	11	22	7.9	20.5	13 188	2.0
Chad	109	43	60.2	5.1	–18 688	12.9
Comoros	12	23	30.6	28.1	7 834	11.2
Congo	25	180	1.1	31.2	115 988	13.0
Congo, Democratic Republic of the	24	258	6.3	50.0	205 787	0.8
Côte d'Ivoire	2 751	511	55.0	14.5	–2 025 365	89.5
Djibouti	11	145	51.8	74.8	82 925	...
Equatorial Guinea	4	30	0.5	6.1	7 900	3.4
Eritrea	2	75	5.6	16.7	61 630	1.8
Ethiopia	330	335	63.1	15.5	59 913	12.9
Gabon	5	164	0.2	16.0	124 602	1.3
Gambia	15	90	54.9	37.9	57 868	13.7

TABLE 6 (cont.)

	Agricultural exports (Million $)	Agricultural imports (Million $)	Agricultural exports as share of total exports (%)	Agricultural imports as share of total imports (%)	Net food imports (Thousand $)	Agricultural exports relative to agricultural GDP (%)
Ghana	728	569	42.1	17.0	–221 118	31.9
Guinea	37	169	4.7	18.0	90 083	4.9
Guinea-Bissau	49	36	79.7	57.4	–20 415	37.5
Kenya	968	460	45.4	12.9	–435 509	51.3
Lesotho	6	105	1.8	13.5	77 886	4.6
Liberia	60	72	12.1	17.9	56 638	25.7
Madagascar	196	114	46.8	15.6	–106 075	14.9
Malawi	406	79	94.7	12.3	–40 469	67.7
Mali	289	167	34.4	20.0	51 387	24.6
Mauritania	35	255	9.1	57.6	152 561	18.8
Mauritius	336	303	18.6	13.8	–82 640	121.4
Mozambique	78	278	8.7	17.9	171 972	8.7
Namibia	193	187	17.2	13.5	36 137	58.1
Niger	67	149	22.9	39.6	41 703	7.4
Nigeria	459	1 899	2.6	19.9	1 122 339	3.2
Rwanda	30	48	42.7	18.1	11 051	4.3
Sao Tome and Principe	5	14	38.9	27.4	3 978	51.8
Senegal	145	593	14.0	30.4	442 822	15.9
Seychelles	1	57	0.3	13.1	30 422	4.0
Sierra Leone	9	137	10.4	28.4	108 593	2.6
Somalia	65	106	69.3	28.9	32 049	...
South Africa	2 569	1 572	8.1	4.8	–747 671	58.6
Saint Helena	...	4	...	46.2	1 969	...
Sudan	348	443	19.2	24.5	214 607	5.8
Swaziland	189	138	19.0	11.6	–72 362	150.2
Tanzania, United Republic of	385	287	51.2	17.9	–44 765	9.5
Togo	106	82	23.2	12.6	26 713	17.9
Uganda	169	147	36.0	11.9	–5 044	9.4
Zambia	111	132	9.6	11.9	37 353	14.2
Zimbabwe	746	181	33.1	10.9	–12 066	36.5
DEVELOPED MARKET ECONOMIES	301 394	310 831	7.1	6.8	8 338 597	85.9
Australia	15 603	3 328	23.3	4.7	–8 432 324	126.2
Austria	4 517	5 342	5.5	6.3	805 735	97.2
Belgium	19 514	17 010	8.9	8.3	–3 406 007	633.6
Canada	17 115	12 988	6.5	5.7	–3 374 826	...
Denmark	10 132	5 166	17.3	10.1	–4 446 778	263.3
Faeroe Islands	15	78	2.6	13.6	59 044	...
Finland	1 182	2 137	2.5	5.9	761 970	27.8
France	36 073	26 381	10.4	7.5	–5 228 084	98.9
Germany	27 594	38 201	4.3	7.2	7 119 670	124.7
Greece	2 635	3 885	23.5	11.4	1 393 752	30.0
Greenland	11	76	2.9	25.7	57 163	...
Iceland	37	214	1.7	8.7	135 285	5.4
Ireland	6 488	3 936	7.4	7.5	–2 264 354	175.1

TABLE 6 (cont.)

	Agricultural exports (Million $)	Agricultural imports (Million $)	Agricultural exports as share of total exports (%)	Agricultural imports as share of total imports (%)	Net food imports (Thousand $)	Agricultural exports relative to agricultural GDP (%)
Israel	1 081	1 927	3.6	5.7	674 233	...
Italy	17 929	23 313	6.8	9.1	3 581 940	59.0
Japan	1 929	35 063	0.5	9.8	23 741 972	3.8
Luxembourg	600	1 209	5.4	8.7	378 813	445.8
Malta	48	283	2.3	9.9	196 626	...
Netherlands	34 087	20 525	13.2	8.9	-7 784 824	316.3
New Zealand	7 096	1 319	46.9	8.5	-5 071 851	...
Norway	484	2 225	0.8	6.2	1 243 922	17.2
Portugal	1 742	4 368	6.2	10.5	2 333 569	41.4
Spain	17 466	13 499	13.1	7.7	-4 520 402	82.2
Saint Pierre and Miquelon	0	5	0.5	8.0	3 124	...
Sweden	2 339	4 823	2.7	6.8	2 019 431	56.6
Switzerland	2 413	5 367	2.7	6.1	1 650 328	...
United Kingdom	15 067	30 347	5.3	8.5	13 917 300	108.8
United States of America	58 199	47 818	8.1	3.9	-7 205 833	37.7
COUNTRIES IN TRANSITION	**19 645**	**30 104**	**5.6**	**8.7**	**7 795 241**	**31.9**
Albania	25	303	7.0	19.4	213 918	2.2
Armenia	58	202	11.9	19.7	141 475	10.2
Azerbaijan	107	260	5.9	13.8	171 437	12.1
Belarus	692	966	8.2	10.1	38 253	49.2
Bosnia and Herzegovina	50	683	6.8	25.2	463 654	6.9
Bulgaria	705	542	11.7	6.3	-61 621	40.2
Croatia	527	967	9.9	8.5	409 293	30.8
Czech Republic	1 475	2 306	3.7	5.4	599 524	57.0
Estonia	391	729	8.4	11.5	207 316	123.1
Georgia	123	242	32.6	29.6	157 239	17.8
Hungary	2 753	1 352	7.7	3.4	-1 295 560	128.3
Kazakhstan	610	545	5.9	7.7	-92 099	30.1
Kyrgyzstan	105	79	20.6	13.3	18 961	18.0
Latvia	268	884	11.0	20.6	451 086	70.6
Lithuania	611	670	10.7	8.5	-63 796	68.3
Macedonia, The former Yugoslav Republic of	215	282	11.1	23.4	151 208	54.7
Moldova, Republic of	427	170	64.0	15.1	-51 545	120.3
Poland	3 320	3 518	7.6	6.1	-443 700	60.3
Romania	504	1 391	3.6	7.3	661 458	8.7
Russian Federation	1 766	9 710	1.5	17.1	5 858 774	9.5
Serbia and Montenegro	466	659	24.0	11.7	59 668	...
Slovakia	544	1 009	3.4	5.7	286 903	52.6
Slovenia	412	772	3.8	6.7	334 070	70.0
Tajikistan	131	114	17.8	15.2	71 674	41.8
Turkmenistan	123	114	4.3	5.0	63 573	12.6
Ukraine	2 355	1 439	12.4	7.6	-602 415	43.1
Uzbekistan	883	196	27.1	6.7	46 491	28.8

TABLE A7
Economic indicators

	Poverty headcount, national	GNP per capita	GDP	GDP per capita	GDP per capita, PPP	Agriculture, value added		Agriculture, value added per worker	
	(% of population)	(Current US$)	(Annual % growth)	(Annual % growth)	(Current international $)	(% of GDP)	(Annual % growth)	(Constant 2000 US$)	(Annual % growth)
	Latest year	2003	1992–2003	1992–2003	2003	2003	1992–2003	2003	1992–2003
WORLD	...	5 578	2.8	3.5	8 368	6.3	2.4	695	2.2
DEVELOPED COUNTRIES	...	21 439	2.3	1.2	23 673	2.4	1.2	5 680	3.1
DEVELOPING COUNTRIES	...	1 351	4.6	4.2	4 307	11.5	3.1	558	2.2
ASIA AND THE PACIFIC	...	1 093	6.6	5.7	4 137	13.7	3.1	423	2.3
Bangladesh	49.8	400	5.0	3.2	1 770	21.8	3.1	313	2.1
Bhutan	...	630	6.7	3.7	...	33.2	3.5	186	1.3
Cambodia	35.9	300	6.5	4.0	2 078	34.5	3.8	300	1.1
China, Hong Kong SAR	...	25 860	3.8	2.3	27 179	...	1.8	...	...
China, Macao SAR	...	...	3.2	1.9	...	...	...	...	...
China, Mainland	4.6	1 100	9.3	8.7	5 003	14.6	3.6	349	2.9
Fiji	...	2 240	3.3	2.2	5 880		0.9	1 966	−0.4
French Polynesia	...	...	2.2	0.4	...	...	...	...	...
India	28.6	540	6.0	4.2	2 892	22.2	3.2	406	1.6
Indonesia	27.1	810	3.8	2.4	3 361	16.6	2.2	547	1.2
Kiribati	...	860	4.3	2.0	...	...	−0.4	...	−0.9
Korea, Republic of	...	12 030	5.5	4.6	17 971	3.2	−0.1	9 792	5.3
Lao People's Dem. Rep.	38.6	340	6.1	3.6	1 759	48.6	4.8	460	2.3
Malaysia	15.5	3 880	5.8	3.4	9 512	9.7	1.4	4 851	2.2
Maldives	...	2 350	7.4	5.0	...	...	...	...	...
Marshall Islands	...	2 710	−0.4	0.6	...	...	...	...	...
Micronesia, Federated States of	...	2 070	1.1	−1.0	...	...	...	...	...
Mongolia	35.6	480	−0.4	−1.6	1 850	28.1	3.9	698	−1.4
Myanmar	...	...	...	6.6	...	...	...	...	...
Nepal	42	240	4.2	1.8	1 420	40.6	2.8	208	0.5
New Caledonia	...	...	1.3	−1.0	...	...	...	...	...
Pakistan	32.6	520	3.7	1.2	2 097	23.3	3.4	695	1.6
Palau	...	6 500	1.6	...	...	...	...	...	...
Papua New Guinea	37.5	500	3.2	0.5	2 619	25.7	3.9	443	1.1
Philippines	36.8	1 080	3.7	1.4	4 321	14.5	2.5	1 040	1.2
Samoa	...	1 440	3.4	2.4	5 854	...	−0.1	1 645	1.2
Singapore	...	21 230	5.5	3.1	24 481	0.1	−3.0	32 073	1.4
Solomon Islands	...	560	0.8	−2.4	1 753	...	...	...	...
Sri Lanka	25	930	4.6	3.3	3 778	19.0	1.5	745	0.4
Thailand	13.1	2 190	4.3	3.5	7 595	9.8	2.3	620	1.9
Timor-Leste	...	460	0.4	−2.3	...	...	−0.8	...	0.5
Tonga	...	1 490	2.4	1.9	6 992	...	1.9	...	4.1
Vanuatu	...	1 180	2.3	−0.6	2 944	...	3.1	...	1.3
Viet Nam	28.9	480	7.4	5.9	2 490	21.8	4.2	296	2.8

TABLE A7 (cont.)

	Poverty headcount, national (% of population) Latest year	GNP per capita (Current US$) 2003	GDP (Annual % growth) 1992–2003	GDP per capita (Annual % growth) 1992–2003	GDP per capita, PPP (Current international $) 2003	Agriculture, value added (% of GDP) 2003	Agriculture, value added (Annual % growth) 1992–2003	Agriculture, value added per worker (Constant 2000 US$) 2003	Agriculture, value added per worker (Annual % growth) 1992–2003
LATIN AMERICA AND THE CARIBBEAN	...	3 323	2.6	1.0	7 493	7.0	2.5	2 966	2.5
Antigua and Barbuda	...	9 160	3.2	1.4	10 294	...	1.3	...	0.0
Argentina	...	3 810	2.3	1.2	12 106	11.1	2.8	9 627	3.0
Bahamas	...	...	1.9	0.1	...	...	...	...	...
Barbados	...	9 260	1.4	1.0	15 720	...	–0.5	18 798	1.2
Belize	...	3 370	5.8	2.7	6 950	...	5.4	...	2.6
Bermuda	...	...	...	1.9	...	...	...	...	...
Bolivia	62.7	900	3.2	1.0	2 587	14.9	2.7	755	0.7
Brazil	22	2 720	2.4	1.0	7 790	5.8	4.0	3 227	5.6
Chile	17	4 360	5.1	3.9	10 274	8.8	2.9	6 341	2.3
Colombia	64	1 810	2.5	0.6	6 702	12.3	–1.4	2 788	–1.6
Costa Rica	22	4 300	4.9	2.9	9 606	8.8	3.4	4 472	2.9
Dominica	...	3 330	0.7	0.8	5 448	...	–1.5	4 659	–0.8
Dominican Republic	28.6	2 130	5.3	3.7	6 823	11.2	3.6	4 142	5.1
Ecuador	35	1 830	2.1	0.3	3 641	7.7	0.1	1 491	–0.9
El Salvador	48.3	2 340	3.8	2.0	4 781	8.5	1.1	1 628	0.3
Grenada	...	3 710	3.0	2.0	7 959	...	–1.3	3 645	2.1
Guatemala	56.2	1 910	3.6	0.9	4 148	22.3	2.4	2 247	0.3
Guyana	...	900	3.5	3.3	4 230	...	5.0	...	5.5
Haiti	...	400	–1.4	–3.2	1 742	...	–4.6	460	–3.9
Honduras	53	970	3.1	0.3	2 665	13.5	3.1	1 223	1.9
Jamaica	18.7	2 980	1.0	0.2	4 104	5.2	1.1	1 957	1.3
Mexico	...	6 230	2.8	1.2	9 168	4.1	2.0	2 866	2.0
Nicaragua	47.9	740	3.5	0.6	3 262	17.9	4.8	1 988	4.9
Panama	37.3	4 060	3.9	2.3	6 854	7.5	4.0	3 605	3.8
Paraguay	21.8	1 110	1.7	–0.7	4 684	27.2	3.3	2 544	1.5
Peru	49	2 140	3.9	2.1	5 260	10.3	4.9	1 770	3.7
Puerto Rico	...	...	4.4	3.6	...	...	...	...	...
Saint Kitts and Nevis	...	6 630	3.7	2.8	12 404	3.0	1.4	2 123	2.5
Saint Lucia	...	4 050	1.8	0.4	5 709	5.4	–3.2	1 738	–5.9
Saint Vincent and the Grenadines	...	3 310	2.2	2.1	6 123	8.7	3.9	2 477	1.7
Suriname	...	2 280	1.7	0.9	...	...	0.3	3 002	0.8
Trinidad and Tobago	21	7 790	4.6	3.6	10 766	1.2	2.4	2 135	2.7
Uruguay	...	3 820	1.3	0.6	8 280	12.8	3.0	7 363	2.9
Venezuela, Bolivarian Republic of	31.3	3 490	–0.2	–2.5	4 919	4.5	0.6	6 071	1.6
NEAR EAST AND NORTH AFRICA	...	2 453	3.5	1.7	5 929	11.2	3.3	2 140	2.1
Algeria	12.2	1 930	2.8	1.0	6 107	10.3	5.5	2 113	1.8
Bahrain	...	...	4.4	1.5	...	...	...	...	...

TABLE A7 (cont.)

	Poverty headcount, national (% of population)	GNP per capita (Current US$)	GDP (Annual % growth)	GDP per capita (Annual % growth)	GDP per capita, PPP (Current international $)	Agriculture, value added (% of GDP)	Agriculture, value added (Annual % growth)	Agriculture, value added per worker (Constant 2000 US$)	Agriculture, value added per worker (Annual % growth)
	Latest year	2003	1992–2003	1992–2003	2003	2003	1992–2003	2003	1992–2003
Cyprus	...	...	4.2	3.4	...	...	...	...	...
Egypt	16.7	1 390	4.3	2.4	3 950	16.1	3.2	1 996	2.2
Iran, Islamic Republic of		2 010	4.1	2.4	6 995	11.3	3.9	2 480	2.4
Jordan	11.7	1 850	4.9	1.6	4 320	2.2	3.0	996	–2.4
Kuwait	...	17 960	5.1	–1.8	18 047	...	...	...	...
Lebanon	...	4 040	3.5	2.0	5 074	12.2	2.1	45 298	5.7
Morocco	19.0	1 310	2.9	1.0	4 004	16.8	12.2	1 711	5.9
Occupied Palestinian Territory	...	1 110	–1.6	–6.8	...	6.2	–6.2	...	...
Oman	...	...	4.2	0.9	...	...	...	...	...
Saudi Arabia	...	9 240	2.3	–0.5	13 226	4.5	1.7	14 618	5.5
Syrian Arab Republic	...	1 160	4.2	1.4	3 576	23.5	4.8	2 768	2.0
Tunisia	7.6	2 240	4.6	3.2	7 161	12.1	3.3	2 639	1.3
Turkey	...	2 800	3.6	1.7	6 772	13.4	1.2	1 766	0.1
United Arab Emirates	...	...	4.3	–2.2	...	...	...	...	...
Yemen	41.8	520	5.5	2.5	889	15.0	6.3	524	3.8
SUB-SAHARAN AFRICA	...	501	2.9	0.5	1 835	16.8	3.7	327	1.4
Angola	...	740	4.0	0.8	2 344	8.8	4.4	161	1.2
Benin	29.0	440	5.0	2.2	1 115	35.7	5.4	606	4.2
Botswana	...	3 530	5.1	2.8	8 714	2.4	–0.8	407	–2.7
Burkina Faso	45.3	300	4.2	1.7	1 174	31.0	3.2	164	0.7
Burundi	36.4	90	–1.2	–3.1	648	49.0	–0.1	101	–1.4
Cameroon	40.2	630	3.0	0.5	2 118	44.2	6.0	1 215	4.9
Cape Verde	...	1 440	5.8	3.2	5 214	6.8	5.4	1 666	4.5
Central African Republic	...	260	1.0	–1.2	1 089	60.8	4.0	425	3.2
Chad	64.0	240	4.5	1.1	1 210	45.6	7.6	257	3.9
Comoros	...	450	1.8	–0.8	1 714	40.9	5.1	386	2.0
Congo	...	650	2.1	–1.1	965	6.2	2.6	347	1.4
Congo, Democratic Republic of the	...	100	–3.6	–5.5	697	...	0.4	...	–1.4
Côte d'Ivoire	...	660	1.6	–1.1	1 476	26.2	2.8	802	2.3
Djibouti	...	910	–0.2	–2.7	2 086	...	1.1	...	–0.6
Equatorial Guinea	...	...	19.7	16.5	...	6.8	5.7	654	3.6
Eritrea	53.0	190	4.8	2.1	849	13.9	6.1	57	0.8
Ethiopia	44.2	90	4.1	1.7	711	41.8	1.3	109	–1.3
Gabon	...	3 340	2.0	–0.7	6 397	8.1	0.8	1 805	1.2
Gambia	57.6	270	3.4	0.1	1 859	30.1	4.9	220	0.9
Ghana	39.5	320	4.3	2.0	2 238	35.8	3.5	346	0.9
Guinea	40.0	430	3.9	1.5	2 097	24.6	4.5	231	2.6
Guinea-Bissau	...	140	1.4	–2.2	711	68.8	3.8	252	1.1
Kenya	52.0	400	1.6	–0.7	1 037	15.8	1.0	148	–1.4

TABLE A7 (cont.)

	Poverty headcount, national	GNP per capita	GDP	GDP per capita	GDP per capita, PPP	Agriculture, value added		Agriculture, value added per worker	
	(% of population)	(Current US$)	(Annual % growth)	(Annual % growth)	(Current international $)	(% of GDP)	(Annual % growth)	(Constant 2000 US$)	(Annual % growth)
	Latest year	2003	1992–2003	1992–2003	2003	2003	1992–2003	2003	1992–2003
Lesotho	...	610	3.4	2.4	2 561	16.6	1.9	499	0.8
Liberia	...	110	9.8	4.2	...	...	...	...	...
Madagascar	71.3	290	2.4	–0.7	809	29.2	1.7	173	–0.6
Malawi	65.3	160	2.7	0.4	605	38.4	8.6	128	6.2
Mali	63.8	290	5.4	2.7	994	38.4	4.8	247	2.6
Mauritania	46.3	400	4.5	1.7	1 766	19.3	2.9	271	0.7
Mauritius	...	4 100	5.0	3.9	11 287	6.1	1.4	4 846	2.5
Mozambique	69.4	210	7.2	4.5	1 117	26.1	6.1	146	3.1
Namibia	...	1 930	3.5	0.9	6 180	10.8	2.6	1 036	1.8
Niger	63.0	200	2.9	–0.5	835	39.9	3.5	174	0.1
Nigeria	34.1	350	3.3	0.5	1 050	26.4	3.6	871	3.2
Rwanda	60.3	220	5.6	1.5	1 268	41.6	6.5	234	2.8
Sao Tome and Principe	...	300	2.5	0.1	...	17.0	3.5	226	1.2
Senegal	33.4	540	4.2	1.4	1 648	16.8	3.2	265	0.6
Seychelles	...	7 490	3.0	1.5	...	3.3	0.5	554	–0.3
Sierra Leone	82.8	150	–2.3	–4.1	548	52.7	–1.7	295	–2.5
South Africa	...	2 750	2.3	0.3	10 346	3.8	2.0	2 251	2.5
Sudan	...	460	5.8	3.3	1 910	...	9.5	...	8.0
Swaziland	40.0	1 350	3.0	0.2	4 726	12.2	0.2	1 189	–0.6
Tanzania, United Rep. of	35.7	300	4.3	1.5	621	45.0	3.6	290	1.4
Togo	32.3	310	3.1	0.1	1 696	40.8	3.3	405	1.2
Uganda	44.0	250	6.6	3.6	1 457	32.4	3.8	231	1.7
Zambia	72.9	380	2.0	–0.3	877	22.8	6.4	210	3.6
Zimbabwe	34.9	...	0.0	–2.2	...	...	2.4	...	1.0
DEVELOPED MARKET ECONOMIES	...	30 147	2.4	1.7	30 767	2.0	1.4	23 081	4.4
Australia	...	21 950	3.8	2.6	29 632	...	2.4	...	2.2
Austria	...	26 810	2.0	1.6	30 094	2.4	3.1	25 117	7.4
Belgium	...	25 760	1.9	1.6	28 335	1.3	3.0	41 876	6.5
Canada	...	24 470	3.3	2.3	30 677	...	0.1	...	2.5
Denmark	...	33 570	2.1	1.7	31 465	2.1	1.6	36 420	6.0
Finland	...	27 060	2.7	2.3	27 619	3.5	2.5	32 031	6.9
France	...	24 730	1.8	1.4	27 677	2.7	1.2	39 038	5.8
Germany	...	25 270	1.3	1.0	27 756	1.1	1.5	22 911	6.4
Greece	...	13 230	2.9	2.1	19 954	6.9	–1.0	9 144	1.1
Iceland	...	30 910	2.9	1.8	31 243	...	0.5	...	1.7
Ireland	...	27 010	7.4	6.2	37 738	...	...	...	...
Israel	...	16 240	3.7	1.2	20 033	...	...	...	...
Italy	...	21 570	1.4	1.3	27 119	2.7	0.1	21 437	4.9
Japan	...	34 180	1.2	0.9	27 967	...	–2.0	...	3.3
Luxembourg	...	45 740	4.4	3.1	62 298	...	2.4	...	...
Malta	...	10 780	3.4	2.7	17 633	...	...	...	...

TABLE A7 (cont.)

	Poverty headcount, national (% of population) Latest year	GNP per capita (Current US$) 2003	GDP (Annual % growth) 1992–2003	GDP per capita (Annual % growth) 1992–2003	GDP per capita, PPP (Current international $) 2003	Agriculture, value added (% of GDP) 2003	Agriculture, value added (Annual % growth) 1992–2003	Agriculture, value added per worker (Constant 2000 US$) 2003	Agriculture, value added per worker (Annual % growth) 1992–2003
Netherlands	...	26 230	2.3	1.6	29 371	...	...	...	4.1
New Zealand	...	15 530	3.4	2.3	22 582	...	2.7	...	2.5
Norway	...	43 400	3.1	2.6	37 670	1.5	1.0	38 043	3.7
Portugal	...	11 800	2.1	1.6	18 126	...	0.6	...	3.4
Spain	...	17 040	2.7	2.2	22 391	...	0.8	...	4.5
Sweden	...	28 910	2.2	1.8	26 750	1.8	0.5	31 960	3.5
Switzerland	...	40 680	1.0	0.4	30 552	...	...	...	...
United Kingdom	...	28 320	2.6	2.4	27 147	...	...	...	...
United States of America	...	37 870	3.2	2.1	37 562	...	4.4	...	6.0
COUNTRIES IN TRANSITION	...	**2 583**	**0.9**	**0.1**	**7 997**	**6.9**	**0.4**	**2 007**	**2.5**
Albania	25.4	1 740	6.1	6.0	4 584	24.7	4.8	1 393	6.6
Armenia	50.9	950	3.7	2.8	3 671	23.5	2.2	2 809	6.2
Azerbaijan	49.0	820	–0.6	–1.4	3 617	14.3	1.3	1 076	0.4
Belarus	41.9	1 600	1.1	1.0	6 052	9.8	–1.4	2 766	2.9
Bosnia and Herzegovina	19.5	1 530	15.8	16.4	5 967	14.9	8.2	...	13.9
Bulgaria	12.8	2 130	0.6	1.2	7 731	11.7	2.9	6 826	8.4
Croatia	...	5 370	2.3	2.2	11 080	8.4	–2.1	9 302	5.3
Czech Republic	...	7 150	1.9	2.0	16 357	3.5	5.0	5 280	3.5
Estonia	8.9	5 380	2.5	2.8	13 539	4.5	–2.0	3 440	1.2
Georgia	11.1	770	–2.6	–2.3	2 588	20.5	1.7	1 503	4.1
Hungary	17.3	6 350	2.8	2.8	14 584	...	0.2	3 990	1.7
Kazakhstan	34.6	1 780	1.3	1.4	6 671	7.8	0.6	1 436	–1.5
Kyrgyzstan	47.6	340	–1.2	–2.0	1 751	38.7	3.3	961	3.5
Latvia	...	4 400	1.9	1.9	10 270	4.5	–3.4	2 513	2.9
Lithuania	...	4 500	0.3	0.4	11 702	7.3	1.6	4 424	6.3
Macedonia, The former Yugoslav Republic of	...	1 980	–0.2	–0.7	6 794	12.2	–1.0	3 096	3.0
Moldova, Republic of	23.3	590	–5.5	–4.4	1 510	22.5	–9.7	706	–4.8
Poland	23.8	5 280	4.1	4.2	11 379	3.1	1.6	1 397	3.1
Romania	21.5	2 260	1.1	1.4	7 277	11.9	0.7	3 621	4.8
Russian Federation	30.9	2 610	–1.2	–0.9	9 230	5.2	–0.3	2 323	2.3
Serbia and Montenegro	...	1 910	2.6	4.8	...	...	...	...	...
Slovakia	...	4 940	2.9	2.6	13 494	3.7	2.8	...	...
Slovenia	...	11 920	3.1	3.1	19 150	...	–0.3	30 713	10.1
Tajikistan	...	210	–5.3	–4.5	1 106	23.4	–1.5	454	1.3
Turkmenistan	...	1 120	5.6	2.2	5 938	...	5.0	1 352	4.1
Ukraine	31.7	970	–4.6	–3.3	5 491	14.1	–2.6	1 400	1.6
Uzbekistan	27.5	420	1.3	–0.4	1 744	35.2	1.7	1 601	2.0

TABLE A8
Total factor productivity

	Total factor productivity change		Efficiency change		Technological change	
	1961–1981	1981–2000	1961–1981	1981–2000	1961–1981	1981–2000
	(Average annual percentage change)					
DEVELOPING COUNTRIES	–2.6	1.7	0.0	–0.4	–2.6	2.0
ASIA AND THE PACIFIC	–3.5	1.9	–0.1	–0.6	–3.4	2.5
Bangladesh	–3.2	1.1	0	0	–3.2	1.1
China, Mainland	–4.4	3.6	0	0	–4.4	3.6
China, Taiwan Province of	0.5	0.3	0	0	0.5	0.3
Fiji	–0.4	–0.3	–0.1	–2.3	–0.2	2
India	–5.2	–1	0	–2.7	–5.2	1.7
Indonesia	–0.5	–1.1	0	0	–0.5	–1.1
Korea, Democratic People's Rep. of	1	1.6	–1.4	1.3	2.5	0.2
Korea, Republic of	–4.5	–1.2	0	0	–4.5	–1.2
Lao People's Dem. Rep.	–0.2	3.3	–0.6	1.9	0.5	1.4
Malaysia	1.8	1.5	0	0	1.8	1.5
Mongolia	–8.3	3.9	–0.7	1.4	–7.7	2.5
Myanmar	0	1.8	0.6	0.5	–0.6	1.3
Nepal	–3.8	1.2	–0.2	0	–3.6	1.2
Pakistan	–0.7	2.7	–1.8	0.2	1.1	2.5
Philippines	1.3	0.4	0	0	1.3	0.4
Sri Lanka	0.7	–0.2	0.2	–1	0.6	0.8
Thailand	0.2	1.4	0.2	0	–0.1	1.4
Viet Nam	0.4	1	–0.2	–0.6	0.7	1.6
LATIN AMERICA AND THE CARIBBEAN	–1.2	0.4	0.1	–0.1	–1.3	0.5
Argentina	–2.2	–3.4	0	0	–2.2	–3.4
Barbados	2.9	0.9	0.3	–1.8	2.6	2.7
Belize	2	1	1.4	–1	0.5	2
Bolivia	0.6	2.6	1	0	–0.4	2.6
Brazil	–3	1.1	0	0	–3	1.1
Chile	1.5	2.9	–0.2	0.1	1.7	2.8
Colombia	1.4	1	0.3	0	1.1	1
Costa Rica	2.6	2.8	1	0.3	1.6	2.4
Cuba	–0.9	0.2	–1.4	–1.6	0.5	1.8
Dominican Republic	0.2	0.5	0	0	0.2	0.5
Ecuador	–1.4	1.3	0	0.1	–1.3	1.2
El Salvador	1.4	–0.1	0.3	–1.3	1.1	1.2
Guadeloupe	–0.6	1.7	–2.4	0.1	1.8	1.6
Guatemala	2.1	0.8	0.7	0	1.4	0.8
Guyana	1.2	1.8	–0.3	0.8	1.5	1
Haiti	–1.4	–0.2	0	0	–1.4	–0.2
Honduras	–1.3	0.4	0.3	–0.6	–1.6	1
Jamaica	0.6	1.6	0.3	–0.8	0.2	2.4
Martinique	–1.5	2.1	–1.4	0	–0.1	2.1
Mexico	1.2	1.1	0.6	–0.6	0.6	1.7
Nicaragua	–4.3	1.5	–1.2	0.7	–3.1	0.9
Panama	–0.2	0.5	–1.1	–0.5	0.9	1
Paraguay	–0.5	–1.9	0	0	–0.5	–1.9

TABLE A8 (cont.)

	Total factor productivity change		Efficiency change		Technological change	
	1961–1981	1981–2000	1961–1981	1981–2000	1961–1981	1981–2000
			(Average annual percentage change)			
Peru	–0.9	2.5	–0.9	0.5	0	2
Saint Lucia	–0.7	–3	0	–2.9	–0.7	–0.2
Saint Vincent and the Grenadines	–1	0.2	–2.9	1.4	1.9	–1.2
Suriname	3.3	–4.3	1.8	–4	1.4	–0.3
Trinidad and Tobago	–1.6	0.5	–0.7	–1.2	–0.9	1.7
Uruguay	–1.5	0.6	0	0	–1.5	0.6
Venezuela, Bolivarian Republic of	1.8	2	1.3	0.1	0.5	1.9
NEAR EAST AND NORTH AFRICA	**0.6**	**2.4**	**–0.2**	**0.2**	**0.7**	**2.1**
Afghanistan	–1.5	2.1	0.3	0	–1.7	2.1
Algeria	–0.8	3.2	–2.2	1.1	1.4	2
Cyprus	3.3	4.4	–0.8	0.4	4.2	4.1
Egypt	1.1	2.1	0	0	1.1	2.1
Iran, Islamic Republic of	0.2	2.3	–0.2	0	0.3	2.3
Iraq	–3.1	–1	–2.3	–1.9	–0.8	0.9
Jordan	–3.4	1.6	–1	–0.1	–2.4	1.7
Lebanon	3.8	2.7	0	0	3.8	2.7
Libyan Arab Jamahiriya	4.6	4.5	3.5	2	1.1	2.4
Morocco	1.7	2.9	0.6	1.2	1.1	1.7
Saudi Arabia	–3.3	4.8	–1.9	2.4	–1.4	2.3
Syrian Arab Republic	1.4	0.3	0	–0.1	1.4	0.4
Tunisia	3.3	2	0.7	2.2	2.5	–0.2
Turkey	1	2.7	0	0	1	2.7
Yemen	–10.3	2.1	–3.3	1.6	–7.3	0.4
SUB-SAHARAN AFRICA	**–3.7**	**1.9**	**0.1**	**0.0**	**–3.8**	**2.0**
Angola	–3.7	5.3	–3.5	4.1	–0.2	1.1
Benin	0.5	2.4	0.5	0.3	0.1	2
Botswana	–2.4	–2.2	–0.2	–1	–2.2	–1.2
Burkina Faso	–9	–0.5	–1	–2.5	–8.1	2
Burundi	–11.5	–0.4	0	0	–11.5	–0.4
Cameroon	–6.8	1.1	0	0	–6.8	1.1
Chad	–3.1	0.2	0	0	–3.1	0.2
Congo	–2.3	–1.4	0	0	–2.3	–1.4
Côte d'Ivoire	–4.1	1.9	0	0	–4.1	1.9
Eritrea	...	–1.9	...	–2.2	...	0.3
Ethiopia	...	3.7	...	0	...	3.7
Gabon	–5.2	2.9	0	0	–5.2	2.9
Gambia	–4.6	–0.7	–2.8	–0.5	–1.9	–0.2
Ghana	–6.6	4.3	0	0	–6.6	4.3
Guinea	–2.4	–1.4	0	0	–2.4	–1.4
Kenya	0.8	1.1	2.1	–0.4	–1.3	1.5
Lesotho	–2.9	–0.5	–2.7	–1.1	–0.2	0.6
Madagascar	–0.9	0.6	0	0	–0.9	0.6
Malawi	–0.8	2.6	–1.3	1.6	0.4	1

TABLE A8 (cont.)

	Total factor productivity change		Efficiency change		Technological change	
	1961–1981	1981–2000	1961–1981	1981–2000	1961–1981	1981–2000
	(Average annual percentage change)					
Mali	−5.2	−1.6	0	−2.2	−5.2	0.6
Mauritius	0.6	−0.3	0	0	0.6	−0.3
Mozambique	−2.3	0.6	0	−0.2	−2.3	0.8
Niger	−6.3	1.3	0	0	−6.3	1.3
Nigeria	−10.5	3.6	0	0	−10.5	3.6
Réunion	2	5.8	−1.1	2.6	3.2	3.1
Rwanda	1.6	0.6	0	0	1.6	0.6
Senegal	−3.4	0.2	−2.3	−0.3	−1.1	0.5
Sierra Leone	−0.6	1.5	−0.7	1.1	0.1	0.4
Sudan	−0.7	2	0	0	−0.7	2
Swaziland	−0.4	1.9	0.1	0.5	−0.5	1.4
Tanzania, United Rep. of	1.1	2.2	1.7	0	−0.6	2.2
Togo	−3.6	1.3	0.4	−0.3	−3.9	1.6
Uganda	1.6	−3.8	0	0	1.6	−3.8
Zambia	−0.4	1.4	−0.1	−1.2	−0.3	2.6
Zimbabwe	0.7	0.8	−0.7	−0.4	1.4	1.3
	1961–1981	1993–2000	1961–1981	1993–2000	1961–1981	1993–2000
COUNTRIES IN TRANSITION	...	1.9	...	0.0	...	1.8
Albania	...	5.8	...	4	...	1.7
Armenia	...	7.5	...	7.3	...	0.2
Azerbaijan	...	8.1	...	6.1	...	1.9
Belarus	...	−1.7	...	−2.4	...	0.7
Bosnia and Herzegovina	...	−3.4	...	−2.8	...	−0.7
Bulgaria	...	4.3	...	1.4	...	2.9
Croatia	...	2.4	...	0	...	2.4
Czech Republic	...	−2	...	0	...	−2
Estonia	...	0.3	...	1.7	...	−1.4
Georgia	...	−0.4	...	−0.9	...	0.5
Hungary	...	0	...	0	...	0
Kazakhstan	...	8.1	...	1.5	...	6.5
Kyrgyzstan	...	3.9	...	1.5	...	2.1
Latvia	...	−0.9	...	0	...	−0.9
Lithuania	...	−2.1	...	−1.3	...	−0.8
Macedonia, The former Yugoslav Republic of	...	−6.9	...	−4.9	...	−2.1
Moldova, Republic of	...	5.7	...	2.9	...	2.8
Poland	...	−0.2	...	0	...	−0.2
Romania	...	0.6	...	−0.9	...	1.5
Russian Federation	...	3.3	...	0	...	3.3
Serbia and Montenegro	...	−1.3	...	0	...	−1.3
Slovakia	...	−2.4	...	−1.7	...	−0.8
Slovenia	...	2.3	...	0	...	2.3
Tajikistan	...	6.1	...	4.2	...	1.8
Turkmenistan	...	0.7	...	−1.5	...	2.2
Ukraine	...	2.8	...	0	...	2.8
Uzbekistan	...	−0.2	...	−1.2	...	1

- **References**

- **Special chapters of**
  The State of Food and Agriculture

- **Selected publications**

References

Aghion, P., Caroli, E. & Garcia-Penalosa, C. 1999. Inequality and growth: the perspective of new growth theories. *Journal of Economic Literature*, 37(4): 1615–1660.

AMAD (Agricultural Marketing Access Database). 2004. Available at http://www.amad.org.

Anderson, K. 2000. Agriculture's "multifunctionality" and the WTO. *The Australian Journal of Agricultural and Resource Economics*, 44(3): 475–494.

Anderson, K. 2002. *Agricultural trade reform and poverty reduction in developing countries*. Policy Discussion Paper 0234. Adelaide, Australia, Centre for International Economic Studies.

Anderson, K. & Martin, W., eds. 2005. Agricultural trade reform and the Doha Development Agenda. *World Economy*, 28(9): 1301–1327.

Anderson, K., Dimaranan, B., Francois, J., Hertel, T., Hoekman, B. & Martin, W. 2001. The cost of rich (and poor) country protection to developing countries. *Journal of African Economies*, 10(3): 227–257.

Anton, J. 2004. *Analysis of the impact of decoupling: overview of on-going OECD work*. Presentation at an Informal Expert Consultation on Domestic Support. FAO, Rome, 30–31 August 2004.

Arndt, C., Jensen, H.T., Robinson, S. & Tarp, F. 2000. Marketing margins and agricultural technology in Mozambique. *Journal of Development Studies*, 37(1): 121–137.

Atkinson, A.B. & Bourguignon, F., eds. 2000. *Handbook of income distribution*. Vol. 1. Amsterdam, North-Holland.

Badiane, O. & Kherallah, M. 1999. Market liberalisation and the poor. *Quarterly Journal of International Agriculture*, 38(4): 341–358.

Bardhan, P. 2004. A powerful, but limited, theory of development. *Economic Development and Cultural Change*, 52(2): 475–486.

Benjamin, D. 1992. Household composition, labor markets, and labor demand: testing for separation in agricultural household models. *Econometrica*, 60(2): 287–322.

Binswanger, H. 1989. The policy response of agriculture. *In* S. Fischer & D. de Tray, eds. *Proceedings of the Annual Conference on Development Economics*, pp. 231–258. Supplement to *The World Bank Economic Review*, Washington, DC, World Bank.

Bruno, M., Ravallion, M. & Squire, L. 1998. Equity and growth in developing countries: old and new perspectives on the policy issues. *In* V. Tanzi & K. Chu, eds. *Income distribution and high quality growth*. Cambridge, Massachusetts, USA, MIT Press.

Carter, C. & Smith, V. 2001. The potential impacts of state trading enterprises on world markets: the exporting country case. *Canadian Journal of Agricultural Economics*, 49(4): 429–439.

Carter, C., Loyns, R. & Berwald, D. 1998. Domestic costs of statutory marketing authorities: the case of the Canadian Wheat Board. *American Journal of Agricultural Economics*, 80(2): 313–324.

Caves, R.E. & Pugel, T.A. 1982. New evidence on competition in grain trade. *Food Research Institute Studies*, 18(3): 261–274.

Chen, S. & Ravallion, M. 2003. Welfare impacts of China's accession to the WTO. *In* D. Bhattasali, S. Li & W. Martin, eds. *China and the WTO: accession, policy reform, and poverty reduction strategies*, pp. 261–282. Washington, DC, World Bank and New York, USA, Oxford University Press.

Chuang, Y.-C. & Lin, C.-M. 1999. Foreign direct investment, R&D and spillover efficiency: evidence from Taiwan's manufacturing firms. *The Journal of Development Studies*, 35(4): 117–137.

Cline, W. 2003. *Trade policy and global poverty*. Washington, DC, Institute for International Economics.

Crow, J.A. 1992. *The epic of Latin America*. Fourth edition. Berkeley, California, USA, The University of California Press.

de Ferranti, D., Perry, G.E., Foster, W., Lederman, D. & Valdés, A. 2005. *Beyond the city: the rural contribution to development*. World Bank Latin American and Caribbean Studies. Washington, DC, World Bank.

de Gorter, H. 2004. *Domestic support disciplines on agriculture in the WTO: where to go from here?* Presentation at an Informal Expert Consultation on Domestic Support. FAO, Rome, 30–31 August 2004.

de Janvry, A., Fafchamps, M. & Sadoulet, E. 1991. Peasant household behavior with missing markets: some paradoxes explained. *The Economic Journal*, 101(409): 1400–1417.

de Janvry, A., Sadoulet, E. & Gordillo de Anda, G. 1995. NAFTA and Mexico's maize producers. *World Development*, 23(8): 1349–1362.

Deininger, K. & Olinto, P. 2000. *Asset distribution, inequality, and growth*. Policy Research Working Paper No. 2375. Washington, DC, World Bank.

Devarajan, S., Go, D.S. & Li, H. 1999. *Quantifying the fiscal effects of trade reform*. Policy Research Working Paper No. 2162. Washington, DC, World Bank.

Dixit, P.M. & Josling, T. 1997. *State trading in agriculture: an analytical framework*. International Agricultural Trade Research Consortium Working Paper 97-4 (available at http://iatrcweb.org/publications/working.htm).

Ebrill, L., Stotsky, J. & Gropp, R. 1999. *Revenue implications of trade liberalization*. Occasional Paper No. 180. Washington, DC, International Monetary Fund.

Edmonds, E. & Pavcnik, N. 2002. *Does globalization increase child labor? Evidence from Vietnam*. Working Paper 8760. Cambridge, Massachusetts, USA, National Bureau of Economic Research.

FAOSTAT (FAO Statistical Databases). 2005. Available at faostat.fao.org.

FAO. 1993. *Design of poverty alleviation in rural areas*, by R. Gaiha. FAO Economic and Social Development Paper 115. Rome

FAO. 2000. *Guidelines for national FIVIMS. Background and principles*. IAWG Guidelines Series No. 1. Rome.

FAO. 2002. *The State of Food Insecurity in the World 2002*. Rome.

FAO. 2003a. *World agriculture: towards 2015/2030. An FAO perspective*, edited by J. Bruinsma. Rome, FAO and London, Earthscan (available at: www.fao.org/docrep/005/y4252e/y4252e00.htm).

FAO. 2003b. *Trade reforms and food security: conceptualizing the linkages*. Rome.

FAO. 2003c. *Anti-Hunger Programme: a twin-track approach to hunger reduction: priorities for national and international action*. Rome (available at ftp://ftp.fao.org/docrep/fao/006/j0563e/j0563e00.pdf).

FAO. 2004a. *The State of Agricultural Commodity Markets 2004*. Rome.

FAO. 2004b. *The State of Food Insecurity in the World 2004*. Rome.

FAO. 2004c. *Socio-economic analysis and policy implications of the roles of agriculture in developing countries research programme summary report 2004*. Rome (available at ftp://ftp.fao.org/es/esa/roa/pdf/summary.pdf).

FAO. 2005a. *Food security in the context of economic and trade policy reforms: insights from country experiences*. CCP 05/11. Rome.

FAO. 2005b. *Domestic support: trade related issues and the empirical evidence*. FAO Trade Policy Technical Notes on issues related to the WTO negotiations on agriculture, No. 5. Rome (available at ftp://ftp.fao.org/docrep/fao/007/j5012e/j5012e00.pdf).

FAO. 2005c. *Export competition: selected issues and the empirical evidence*. FAO Trade Policy Technical Notes on issues related to the WTO negotiations on agriculture, No. 4. Rome (available at ftp://ftp.fao.org/docrep/fao/007/j5013e/j5013e00.pdf).

Francois, J.F. 2001a. *Modeling the impact of WTO negotiations on EU agriculture: an application of the GTAP model*. Paper prepared for the European Commission-FAIR Project "Assessment of the GTAP modeling framework for policy analysis from a European Perspective".

Francois, J.F. 2001b. *The next WTO round: North-South stakes in new market access negotiations*. Adelaide, Australia, Centre for International Economic Studies and Amsterdam and Rotterdam, Netherlands, Tinbergen Institute.

Francois, J.F. & Martin, W. 2004. Commercial policy variability, bindings, and market access. *European Economic Review*, 48(3): 665–679.

Francois, J.F., van Meijl, H. & van Tongeren, F.W. 2003. *Economic benefits of the Doha Round for the Netherlands*. Report submitted to the Ministry of Economic Affairs, Directorate-General for Foreign Economic Relations, Netherlands. The Hague, Agricultural Economics Research Institute.

Francois, J.F., van Meijl, H. & van Tongeren, F.W. 2005. Trade liberalization in the Doha Development Round. *Economic Policy*, 20(42): 349–391.

Friedman, J. & Levinsohn, J. 2002. The distributional impacts of Indonesia's financial crisis on household welfare: a "rapid response" methodology. *The World Bank Economic Review*, 16(3): 397–423.

GATT. 1994. *The results of the Uruguay Round of multilateral trade negotiations: legal texts*. Geneva, Switzerland, GATT Secretariat.

Gisselquist, D. & Pray, C. 1997. The impact of Turkey's 1980s seed regulatory reform. *In* D. Gisselquist & J. Srivastava, eds. *Easing barriers to movement of plant varieties for agricultural development*, pp. 113–131. Discussion Paper 367. Washington, DC, World Bank.

GTAP (Global Trade Analysis Project) 6.4 database. Available at http://www.gtap.agecon.purdue.edu/databases/v6/default.asp.

Haley, S. 1995. *U.S. imports of Canadian wheat: estimating the effect of the U.S. Export Enhancement Program.* International Agricultural Trade Research Consortium Working Paper 95-2 (available at http://iatrcweb.org/publications/working.htm).

Hamilton, S.F. & Stiegert, K.W. 2002. An empirical test of the rent-shifting hypothesis: the case of state trading enterprises. *Journal of International Economics*, 58(1): 135–157.

Harrison, G.W., Rutherford, T.F. & Tarr, D.G. 1997. Quantifying the Uruguay Round. *The Economic Journal*, 107: 1405–1430.

Hathaway, D.E. & Ingco, M.D. 1996. Agricultural liberalization and the Uruguay Round. In W. Martin & L.A. Winters, eds. *The Uruguay Round and the developing countries*, pp. 30–58. Cambridge, UK, Cambridge University Press.

Hayami, Y. and Ruttan, V.W. 1985. *Agricultural development: an international perspective*. 2nd edition. Baltimore, USA, Johns Hopkins University Press.

Hertel, T.W. & Ivanic, M. 2005. *Agricultural trade policy and poverty in developing countries*. ESA Working Paper. Rome, FAO.

Hertel, T.W. & Reimer, J.J. 2004. Predicting the poverty impacts of trade liberalization: a survey. Unpublished manuscript. West Lafayette, Indiana, USA, Center for Global Trade Analysis, Purdue University.

Hertel, T.W. & Winters, L.A., eds. 2005 (forthcoming). *Putting development back into the Doha Agenda: poverty impacts of a WTO agreement*. Washington, DC, World Bank and New York, USA, Oxford University Press.

Hertel, T.W., Zhai, F. & Wang, Z. 2004. Implications of WTO accession for poverty in China. In D. Bhattasali, S. Li & W. Martin, eds. *China and the WTO: accession, policy reform, and poverty reduction strategies*, pp. 283–303. Washington, DC, World Bank and New York, USA, Oxford University Press.

Hertel, T.W., Ivanic, M., Preckel, P.V. & Cranfield, J.A.L. 2004. The earnings effects of multilateral trade liberalization: implications for poverty. *The World Bank Economic Review*, 18(2): 205–236.

Hertel, T.W., Ivanic, M., Preckel, P.V., Cranfield, J.A.L. & Martin, W. 2003. Short- versus long-run implications of trade liberalization for poverty in three developing countries. *American Journal of Agricultural Economics*, 85(5): 1299–1306.

IMF (International Monetary Fund)/World Bank. 2002. *Market access for developing countries exports – selected issues*. Washington, DC, World Bank. (mimeo)

Ingco, M. & Nash, J.D., eds. 2004. *Agriculture and the WTO: creating a trading system for development*. Washington, DC, World Bank and New York, USA, Oxford University Press.

Jales, M. 2004. *The impact of export competition policies: export subsidies and export credits*. Presentation at an Informal Expert Consultation on Export Competition: Equivalence of Alternative Policies and Mechanisms. FAO, Rome, 25–26 November 2004.

Kehoe P.J. & Kehoe, T.J. 1994. A primer on static applied general equilibrium models. *Federal Reserve Bank of Minneapolis Quarterly Review*, 18(2) (available at http://minneapolisfed.org/research/QR/QR1821.pdf).

Khan, H.A. 2003. *Innovation and growth in East Asia: the future of miracles*. Houndmills, Basingstoke, UK, Macmillan.

Larue, B., Fulton, M. & Veeman, M. 1999. On exporting by import state traders and peculiar effects of negotiated minimum access commitments. *Canadian Journal of Agricultural Economics*, 47(4): 375–384.

Lipton, M. & Ravallion, M. 1995. Poverty and policy. In J. Behrman & T.N. Srinivasen, eds. *Handbook of development economics*, Vol. 3B, Chapter 41, pp. 2551–2657. Amsterdam, North-Holland.

Malmquist, S. 1953. Index numbers and indifference surfaces. *Trabajos de Estatistica*, 4: 209–242.

Martin, W. & Zhi, W. 2005 (forthcoming). *The landscape of world agricultural protection*. ESA Working Paper. Rome, FAO.

McCorriston, S. & MacLaren, D. 2002. Perspectives on the state trading issue in the WTO negotiations. *European Review of Agricultural Economics*, 29(1): 131–154.

McCorriston, S. & MacLaren, D. 2004. *Trade distorting STEs*. Presentation at an Informal Expert Consultation on Export Competition: Equivalence of Alternative Policies and Mechanisms. FAO, Rome, 25–26 November 2004.

McCorriston, S. & MacLaren, D. 2005. Domestic market structure and trade: modelling the effects of trade distorting state trading enterprises. In S. Jayasuriya, ed. *Trade theory and analytical models: essays in honour of Peter Lloyd*. Vol. 1. Cheltenham, UK, Edward Edgar.

McCulloch, N., Winters, L. & Cirera, X. 2001. *Trade liberalisation and poverty: a handbook*.

London, Centre for Economic and Policy Research and UK Department for International Development.

McMillan, M., Rodrik, D. & Welch, H.W. 2002. *When economic reform goes wrong: cashews in Mozambique.* Faculty Research Working Paper Series, Cambridge, Massachusetts, USA, Harvard University.

Messerlin, P. 2003. *Agriculture in the Doha Agenda.* Policy Research Working Paper 3009. Washington, DC, World Bank.

Milner, C., Morrissey, O. & Rudaheranwa, N. 2001. Policy and non-policy barriers to trade and implicit taxation of exports in Uganda. *Journal of Development Studies*, 37(2): 67–90.

Minot, N.W. 1998. Distributional and nutritional impact of devaluation in Rwanda. *Economic Development & Cultural Change*, 46(2): 379–403.

Minot, N.W. & Goletti, F. 2000. *Rice market liberalisation and poverty in Vietnam.* Research Report 114. Washington, DC, International Food Policy Research Institute.

Morrison, J.A. 2002 *The differential impacts of trade liberalisation on food security: a research agenda.* Paper presented at the FAO Expert Consultation on Trade and Food Security: Conceptualising the linkages, Rome, 11–12 July 2002.

Nicita, A. 2004. *Who benefited from trade liberalization in Mexico? Measuring the effects on household welfare.* Policy Research Working Paper 3265. Washington, DC, World Bank.

OECD. 2000a. *An analysis of officially supported export credits in agriculture.* COM/AGR/TD/WP(2000)91/FINAL. Paris.

OECD. 2000b. *A review of state trading enterprises in agriculture in OECD member countries: an inventory.* COM/AGR/APM/TD/WP(2000)19. Paris.

OECD. 2000c. *A review of state trading enterprises in agriculture in OECD member countries.* COM/AGR/APM/TD/WP(2000)18. Paris.

OECD. 2004. *Risk effects of PSE crop measures.* AGR/CA/APM(2002)13/FINAL. Paris.

OECD. 2005. *Producer and Consumer Support Estimates,* OECD Database 1986–2004. Available at www.oecd.org.

Porto, G.G. 2003a. *Trade reforms, market access and poverty in Argentina.* Washington, DC, World Bank.

Porto, G.G. 2003b. *Using survey data to assess the distributional effects of trade policy.* Washington, DC, World Bank.

Rae, A. & Josling, T. 2003. Processed food trade and developing countries: protection and trade liberalization. *Food Policy*, 28(2): 147–166.

Ravallion, M. 1990. Rural welfare effects of food price changes under induced wage responses: theory and evidence for Bangladesh. *Oxford Economic Papers*, 42: 574–585.

Ravallion, M. & Datt, G. 1996. How important to India's poor is the sectoral composition of economic growth? *The World Bank Economic Review*, 10(1): 1–25.

Ravallion, M. & Datt, G. 1999. *When is growth pro-poor? Evidence from the diverse experiences of India's states.* Policy Research Working Paper 2263. Washington, DC, World Bank.

Ravallion, M. & Lokshin, M. 2004. *Gainers and losers from trade reform in Morocco.* Policy Research Working Paper 3368. Washington, DC, World Bank.

Reardon, T. & Berdegué, J.A. 2002. The rapid rise of supermarkets in Latin America: challenges and opportunities for development. *Development Policy Review*, 20(4): 371–388.

Rodríguez, F. & Rodrik, D. 1999. *Trade policy and economic growth: a sceptic's guide to the cross-national evidence.* Discussion Paper 2143, London, Centre for Economic Policy Research.

Roland-Holst, D. 2004. *CGE methods for poverty incidence analysis: an application to Vietnam's WTO accession.* Paper presented at the Seventh Annual Conference on Global Economic Analysis: Trade, Poverty, and the Environment, World Bank, Washington, DC, 17–19 June 2004.

Romer, P. 1994. New goods, old theory and the welfare cost of trade restrictions. *Journal of Development Economics*, 43: 5–38.

Runge, C. Ford, Senauer, B., Pardey, P.G. & Rosegrant, M.W. 2003. *Ending hunger in our lifetime: food security and globalization.* Washington, DC, International Food Policy Research Institute and Baltimore, Maryland, USA, The Johns Hopkins University Press.

Sadoulet, E. & de Janvry, A. 1995. *Quantitative development policy analysis.* Baltimore, Maryland, USA, The Johns Hopkins University Press.

Sarris, A. 2003. *The role of agriculture in economic development and poverty reduction.* Washington, DC, World Bank. (mimeo)

Schiff, M. & Valdés, A. 1998. The plundering of agriculture in developing countries. In C.K. Eicher & J. Schaatz, eds. *International agricultural development.* Third edition, pp. 226–233. Baltimore, Maryland, USA and London, The Johns Hopkins University Press.

Scoppola, M. 2004. *Determining appropriate disciplines for STE: STE and private exporters.* Presentation at an Informal Expert Consultation on Export Competition: Equivalence of Alternative Policies and Mechanisms. FAO, Rome, 25–26 November 2004.

Sicular, T. & Zhao, Y. 2002. *Employment, earnings and the rural poverty impacts of China's WTO Accession.* Paper presented at the Democratic Republic of the Congo/World Bank Workshop on WTO Accession and Poverty, Beijing, May.

Smith, J.P., Thomas, D., Frankenberg, E., Beegle, K. & Teruel, G. 2002. Wages, employment and economic shocks: evidence from Indonesia. *Journal of Population Economics,* 15: 161–193.

Sumner, D. & Boltuck, R. 2001. *Anatomy of the global wheat market and the role of the Canadian Wheat Board.* Winnipeg, Canada, Canadian Wheat Board.

Tangermann, S. 1998. Implementation of the Uruguay Round Agreement on Agriculture by major developed countries. *In* H. Thomas & J. Whalley, eds. *Uruguay Round results and the emerging trade agenda.* New York, USA, and Geneva, Switzerland, United Nations Conference on Trade and Development.

Taylor, J., Yunez-Naude, A. & Dyer, G. 2003. *Disaggregated impacts of policy reform: description of a case study using data from the Mexico National Rural Household Survey.* Paper presented at the OECD Global Forum on Agriculture, Paris, 10–11 December 2003.

Thomas, D., Frankenberg, E., Beegle, K. & Teruel, G. 1999. *Household budgets, household composition and the crisis in Indonesia: evidence from Longitudinal Household Survey data.* Paper presented at the Annual Meeting of the 1999 Population Association of America, New York, USA, 25–27 March 1999.

Timmer, C.P. 1995. Getting agriculture moving: do markets provide the right signals? *Food Policy,* 20(5): 455–472.

UNCTAD (United Nations Conference on Trade and Development). 2004. *The Least Developed Countries Report 2004: linking international trade with poverty reduction.* New York, USA, and Geneva, Switzerland (also available at http://www.unctad.org/en/docs/ldc2004_en.pdf).

UNDP (United Nations Development Programme). 2003. *Making global trade work for people.* London, UK and Sterling, USA, Earthscan Publications Ltd (also available at http://www.undp.org/mdg/globaltrade.pdf).

USDA (United States Department of Agriculture). 2001. *The road ahead: agricultural policy reform in the WTO. Summary report,* edited by M.E. Burfisher. Agricultural Economic Report No. 797. Washington, DC, Economic Research Service, USDA.

Valdés A. & Foster, W. 2003. *The positive externalities of Chilean agriculture: the significance of its growth and export orientation. A Synthesis of the ROA Chile Case Study.* Rome, FAO.

van Tongeren, F. 2005 (forthcoming). *Macroeconomic implications of agricultural trade policy.* ESA Working Paper. Rome, FAO.

Vogel, S.J. 1994. Structural changes in agriculture: production linkages and agricultural demand-led industrialization. *Oxford Economic Papers,* 46(1): 136–157.

Wainio, J. & Gibson, P. 2004. Measuring agricultural tariff protection. *In* USDA. *U.S. agriculture and the free trade area of the Americas* (ed. M.E. Burfisher), pp. 52–66. Agricultural Economic Report No. 827. Washington, DC, Economic Research Service, USDA.

Wainio, J., Gibson, P. & Whitley, D. 2001. Options for reducing agricultural tariffs. *In* USDA. *Agricultural policy reform: the road ahead,* edited by M.E. Burfisher, pp. 43–57. Agricultural Economic Report No. 802. Washington, DC, Economic Research Service, USDA.

Walmsley, T., Hertel, T. & Ianchovichina, E. 2005 (forthcoming). Assessing the impact of China's WTO accession on investment. *Pacific Economic Review.*

Winters, L.A. 2002. Trade liberalisation and poverty: what are the links? *The World Economy,* 25(9): 1339–1367.

Winters, L.A., McCulloch, N. & McKay, A. 2004. Trade liberalization and poverty: the evidence so far. *Journal of Economic Literature,* 42(1): 72–115.

Woolf, G., ed. 2003. *The Cambridge illustrated history of the Roman world.* Cambridge, Massachusetts, USA, Cambridge University Press.

World Bank. 2003. *Global Economic Prospects 2004: realizing the development promise of the Doha Agenda.* Washington, DC (also available at http://siteresources.worldbank.org/INTRGEP2004/Resources/gep2004fulltext.pdf).

World Bank. 2005a. *Global agricultural trade and developing countries,* edited by M.A. Aksoy & J.C. Beghin. Washington, DC.

World Bank. 2005b. *Global Economic Prospects 2005: trade, regionalism, and development.* Washington, DC (also available at http://siteresources.worldbank.org/INTGEP2005/Resources/gep2005.pdf).

World Bank. 2005c. *WDI Online.* World Development Indicators database (available at http://publications.worldbank.org/WDI/).

WTO (World Trade Organization). 2003. *World Trade Report 2003.* Geneva, Switzerland (also available at http://www.wto.org/english/res_e/booksp_e/anrep_e/world_trade_report_2003_e.pdf).

WTO (World Trade Organization). 2004a. *World Trade Report 2004: exploring the linkage between the domestic policy environment and international trade.* Geneva, Switzerland (also available at http://www.wto.org/english/res_e/booksp_e/anrep_e/world_trade_report04_e.pdf).

WTO (World Trade Organization). 2004b. *Doha Work Programme, Decision Adopted by the General Council on 1 August 2004.* WT/L/579. Geneva, Switzerland.

Young, L. 2004a. *State trading enterprises: possible disciplines.* Presentation at an Informal Expert Consultation on Export Competition: Equivalence of Alternative Policies and Mechanisms. FAO, Rome, 25–26 November 2004.

Young, L. 2004b. *Food aid: possible disciplines.* Presentation at an Informal Expert Consultation on Export Competition: Equivalence of Alternative Policies and Mechanisms. FAO, Rome, 25–26 November 2004.

Young, C.E. & Westcott, P.C. 2000. How decoupled is U.S. agricultural support for major crops? *American Journal of Agricultural Economics*, 82(3): 762–767.

Special chapters of
The State of Food and Agriculture

In addition to the usual review of the recent world food and agricultural situation, each issue of this report since 1957 has included one or more special studies on problems of longer-term interest. Special chapters in earlier issues have covered the following subjects:

1957 Factors influencing the trend of food consumption
Postwar changes in some institutional factors affecting agriculture

1958 Food and agricultural developments in Africa south of the Sahara
The growth of forest industries and their impact on the world's forests

1959 Agricultural incomes and levels of living in countries at different stages of economic development
Some general problems of agricultural development in less-developed countries in the light of postwar experience

1960 Programming for agricultural development

1961 Land reform and institutional change
Agricultural extension, education and research in Africa, Asia and Latin America

1962 The role of forest industries in the attack on economic underdevelopment
The livestock industry in less-developed countries

1963 Basic factors affecting the growth of productivity in agriculture
Fertilizer use: spearhead of agricultural development

1964 Protein nutrition: needs and prospects
Synthetics and their effects on agricultural trade

1966 Agriculture and industrialization
Rice in the world food economy

1967 Incentives and disincentives for farmers in developing countries
The management of fishery resources

1968 Raising agricultural productivity in developing countries through technological improvement
Improved storage and its contribution to world food supplies

1969 Agricultural marketing improvement programmes: some lessons from recent experience
Modernizing institutions to promote forestry development

1970 Agriculture at the threshold of the Second Development Decade

1971 Water pollution and its effects on living aquatic resources and fisheries

Year	Title
1972	Education and training for development
	Accelerating agricultural research in the developing countries
1973	Agricultural employment in developing countries
1974	Population, food supply and agricultural development
1975	The Second United Nations Development Decade: mid-term review and appraisal
1976	Energy and agriculture
1977	The state of natural resources and the human environment for food and agriculture
1978	Problems and strategies in developing regions
1979	Forestry and rural development
1980	Marine fisheries in the new era of national jurisdiction
1981	Rural poverty in developing countries and means of poverty alleviation
1982	Livestock production: a world perspective
1983	Women in developing agriculture
1984	Urbanization, agriculture and food systems
1985	Energy use in agricultural production
	Environmental trends in food and agriculture
	Agricultural marketing and development
1986	Financing agricultural development
1987–88	Changing priorities for agricultural science and technology in developing countries
1989	Sustainable development and natural resource management
1990	Structural adjustment and agriculture
1991	Agricultural policies and issues: lessons from the 1980s and prospects for the 1990s
1992	Marine fisheries and the law of the sea: a decade of change
1993	Water policies and agriculture
1994	Forest development and policy dilemmas
1995	Agricultural trade: entering a new era?
1996	Food security: some macroeconomic dimensions
1997	The agroprocessing industry and economic development
1998	Rural non-farm income in developing countries
2000	World food and agriculture: lessons from the past 50 years
2001	Economic impacts of transboundary plant pests and animal diseases
2002	Agriculture and global public goods ten years after the Earth Summit
2003–04	Agricultural biotechnology: meeting the needs of the poor?

Selected publications

FAO FLAGSHIP PUBLICATIONS

(available at www.fao.org/sof)

The State of Food and Agriculture
The State of Agricultural Commodities
The State of Food Insecurity in the World
The State of World Fisheries and Aquaculture
State of the World's Forests

AGRICULTURAL AND DEVELOPMENT ECONOMICS DIVISION (ESA) PUBLICATIONS

(available at www.fao.org/es/esa)

BOOKS AND JOURNAL ARTICLES

Moving away from poverty:
a spatial analysis of poverty and migration in Albania
Journal of Southern Europe and the Balkans, 7(2): 175–193
(A. Zezza, G. Carletto and B. Davis, August 2005)

Agricultural biotechnology for developing countries:
an FAO perspective
Farm Policy Journal, 1(2): 4–12
(T. Raney, August 2004)

The state of global food insecurity and the benefits
of hunger reduction
Fome Zero: políticas públicas e cidadania.
Cadernos do CEAM, Vol. 4, No. 14
(Brasilia: Universidade de Brasilia)
(H. de Haen and B. Davis, 2004)

Hogares, pobreza y políticas en épocas de crisis.
México, 1992–1996
Revista de la CEPAL, 82: 191–212
(B. Davis, S. Handa, and H. Soto, April, 2004)

Sustaining food security in the developing world:
the top five policy challenges
Quarterly Journal of International Agriculture, 42(3): 261–272
(P. Pingali, 2003)

Agricultural biodiversity and biotechnology in economic development
(New York, USA: Springer)
(J. Cooper, L.M. Lipper and D. Zilberman)

ESA WORKING PAPERS
05-06   *Measuring technical efficiency of wheat farmers in Egypt*
        (A. Croppenstedt)

05-05	*Food aid: a primer* (S. Lowder and T. Raney)
05-04	*Transaction costs, institutions and smallholder market integration: potato producers in Peru* (I. Maltsoglou and A. Tanyeri-Abur)
05-03	*Familiar faces, familiar places: the role of family networks and previous experience for Albanian migrants* (G. Carletto, B. Davis and M. Stampini)
05-02	*Moving away from poverty: a spatial analysis of poverty and migration in Albania* (A. Zezza, G. Carletto and B. Davis)
05-01	*Monitoring poverty without consumption data: an application using the Albania panel survey* (C. Azzarri, G. Carletto, B. Davis and A. Zezza)
04-22	*Investing in agriculture for growth and food security in the ACP countries* (J. Skoet, K. Stamoulis and A. Deuss)
04-21	*Estimating poverty over time and space: construction of a time-variant poverty index for Costa Rica* (R. Cavatassi, B. Davis and L. Lipper)
04-20	*Will buying tropical forest carbon benefit the poor? Evidence from Costa Rica* (S. Kerr, A. Pfaff, R. Cavatassi, B. Davis, L. Lipper, A. Sanchez and J. Hendy)
04-19	*Effects of poverty on deforestation: distinguishing behaviour from location* (S. Kerr, A. Pfaff, R. Cavatassi, B. Davis, L. Lipper, A. Sanchez and J. Timmins)
04-18	*Understanding vulnerability to food insecurity: lessons from vulnerable livelihood profiling* (C. Løvendal, M. Knowles and N. Horii)
04-17	*Westernization of Asian diets and the transformation of food systems: implications for research and policy* (P. Pingali)
04-16	*Identifying the factors that influence small-scale farmers' transaction costs in relation to seed acquisition* (L.B. Badstue)
04-15	*Poverty, livestock and household typologies in Nepal* (I. Maltsoglou and K. Taniguchi)
04-14	*National agricultural biotechnology research capacity in developing countries* (J. Cohen, J. Komen and J. Falck Zepeda)
04-13	*Internal mobility and international migration in Albania* (G. Carletto, B. Davis, M. Stampini, S. Trento and A. Zezza)
04-12	*Being poor, feeling poorer: combining objective and subjective measures of welfare in Albania* (G. Carletto and A. Zezza)
04-11	*Food insecurity and vulnerability in Viet Nam: profiles of four vulnerable groups* (FAO Food Security and Agricultural Projects Analysis Service)
04-10	*Food insecurity and vulnerability in Nepal: profiles of seven vulnerable groups* (FAO Food Security and Agricultural Projects Analysis Service)

04-09	*Public attitudes towards agricultural biotechnology* (T. J. Hoban)
04-08	*The economic impacts of biotechnology-based technological innovations* (G. Traxler)
04-07	*Private research and public goods: implications of biotechnology for biodiversity* (T. Raney and P. Pingali)
04-06	*Interactions between the agricultural sector and the HIV/AIDS pandemic: implications for agricultural policy* (T. S. Jayne, M. Villarreal, P. Pingali and G. Hemrich)
04-05	*Globalization of Indian diets and the transformation of food supply systems* (P. Pingali and Y. Khwaja)
04-04	*Agricultural policy indicators* (T. Josling and A. Valdés)
04-03	*Resource abundance, poverty and development* (E.H. Bulte, R. Damania and R.T. Deacon)
04-02	*Conflicts, rural development and food security in West Africa* (M. Flores)
04-01	*Valuation methods for environmental benefits in forestry and watershed investment projects* (R. Cavatassi)
03-22	*Linkages and rural non-farm employment creation: changing challenges and policies in Indonesia* (S. Kristiansen)
03-21	*Information asymmetry and economic concentration: the case of hens and eggs in eastern Indonesia* (S. Kristiansen)
03-20	*Do futures benefit farmers who adopt them?* (S.H. Lence)
03-19	*The economics of food safety in developing countries* (S. Henson)
03-18	*Food security and agriculture in the low income food deficit countries: 10 years after the Uruguay Round* (P. Pingali and R. Stringer)
03-17	*A conceptual framework for national agricultural, rural development, and food security strategies and policies* (K.G. Stamoulis and A. Zezza)
03-16	*Can public transfers reduce Mexican migration? A study based on randomized experimental data* (G. Stecklov, P. Winters, M. Stampini and B. Davis)
03-15	*Diversification in South Asian agriculture: trends and constraints* (K. Dorjee, S. Broca and P. Pingali)
03-14	*Determinants of cereal diversity in communities and on household farms of the northern Ethiopian Highlands* (S. Benin, B. Gebremedhin, M. Smale, J. Pender and S. Ehui)
03-13	*Land use change, carbon sequestration and poverty alleviation* (L. Lipper and R. Cavatassi)
03-12	*Social capital and poverty lessons from case studies in Mexico and Central America* (M. Flores and F. Rello)

Further and Higher Education Act 1992

CHAPTER 13

ARRANGEMENT OF SECTIONS

PART I
FURTHER EDUCATION
CHAPTER I
RESPONSIBILITY FOR FURTHER EDUCATION

The new funding councils

Section
1. The Further Education Funding Councils.

The new further education sector

2. Full-time education for 16 to 18 year-olds.
3. Part-time education, and full-time education for those over 18.
4. Persons with learning difficulties.

Finance

5. Administration of funds by councils.
6. Administration of funds: supplementary.
7. Grants to councils.

Further functions

8. Supplementary functions.
9. Assessment of quality of education provided by institutions.

Adjustment of local education authority sector

10. Functions of local education authorities in respect of secondary education.
11. Functions of local education authorities in respect of further education.

Provision of further education in schools

12. Provision of further education in maintained schools.

Section
13. Provision of further education in grant-maintained schools.

General

14. Meaning of "further education", "secondary education", "school" and "pupil".

CHAPTER II

INSTITUTIONS WITHIN THE FURTHER EDUCATION SECTOR

The further education corporations

15. Initial incorporation of existing institutions.
16. Orders incorporating further institutions.
17. "Further education corporation" and "operative date".
18. Principal powers of a further education corporation.
19. Supplementary powers of a further education corporation.
20. Constitution of corporation and conduct of the institution.
21. Initial instruments and articles.
22. Subsequent instruments and articles.

Transfer of property, etc., to further education corporations

23. Transfer of property, etc.: institutions maintained by local education authorities.
24. Provisions supplementary to section 23.
25. Transfer of property, etc.: grant-maintained schools.
26. Transfer of staff to further education corporations.

Dissolution of further education corporations

27. Dissolution of further education corporations.

Designation of institutions for funding by the councils

28. Designation of institutions.
29. Government and conduct of designated institutions.
30. Special provision for voluntary aided sixth form colleges.
31. Designated institutions conducted by companies.
32. Transfer of property, etc., to designated institutions.
33. Provisions supplementary to section 32.

Property, rights and liabilities: general

34. Making additional property available for use.
35. Voluntary transfers of staff in connection with section 34.
36. General provisions about transfers under Chapter II.
37. Attribution of surpluses and deficits.
38. Payments by council in respect of loan liabilities.
39. Control of disposals of land.
40. Wrongful disposals of land.
41. Control of contracts.
42. Wrongful contracts.
43. Remuneration of employees.

Miscellaneous

44. Collective worship.